SUN-DRIED TOMATOES

ANDREA CHESMAN

THE CROSSING PRESS SPECIALTY COOKBOOKS

THE CROSSING PRESS

FREEDOM, CALIFORNIA

Cote D'Azur Fusilli Salad, Tuna Pasta Salad, and Antipasto Pasta Salad first appeared in Pasta Salads! by Susan Janine Meyer (The Crossing Press, 1986). Red Basil Pesto and Vermicelli with Red Basil Pesto first appeared in Pestos! by Dorothy Rankin (The Crossing Press, 1985). The author gratefully acknowledges their use. The recipe for Deep-Dish Kale and Cheese Pizza first appeared in Vermont Life (Autumn, 1986).

Cover photo courtesy of Sonoma Dried Tomatoes
Cover design by Victoria May
Text design by Sheryl Karas
Printed in the U.S.A.

Library of Congress Cataloging-in-Publication Data
Chesman, Andrea
Sun-dried tomatoes / Andrea Chesman.
p. cm. — (The Crossing Press Specialty cookbooks)
Includes index.
ISBN 0-89594-900-8 (paper)
1. Cookery (Tomatoes) 2. Dried tomatoes. I Title. II. Series.
TX803.T6C49 1997
641.6'5642--dc21

97-24473
CIP

CONTENTS

The author wishes to thank Timber Crest Farms in Healdsburg, California; Genovesi Food Company in Dayton, Ohio, L'Esprit de Campagne in Winchester, Virginia; Doug Mack of Mary's Restaurant in Bristol, Vermont; and Scott Vineberg of Francesca's Restaurant in Shelburne, Vermont, for inspiration and help with recipe development.

ABOUT SUN-DRIED TOMATOES

I still remember the first time I tasted a sun-dried tomato. It was a revelation. It tasted like a tomato, only more so. The flavor exploded in my mouth, a heady combination of salt and sweet, with a bass note of earthiness and sun. Quite simply, that sun-dried tomato was unlike anything I had ever tasted.

At the time I was writing a regular column on food for a newspaper in Burlington, Vermont, and I noticed that sun-dried tomatoes were popping up on menus all over town. Justifying my culinary curiosity under the guise of journalistic inquiry, I decided to write an article about these tasty items. I collected recipes from local restaurateurs and tasted various brands of sun-dried tomatoes, some packed in oil, some not.

Some tomatoes replayed for me that intensely flavorful experience, sending me to the kitchen to find more and more uses for them. But I also found brands that were unspeakably salty, and some dried tomatoes so tough they did not soften in hot water. The price variations were significant, too. Clearly, not all dried tomatoes were equal.

Eventually I concluded I could dry my own tomatoes very nicely, save money, and have a virtually endless supply of a gourmet ingredient. From there it was easy to start using sun-dried tomatoes in all sorts of combinations. This book is the culmination of my experiments.

There are two ways to look at a sun-dried tomato—as a replacement for a vine-ripened fresh tomato and as a uniquely flavored ingredient on its own. As a replacement for fresh tomatoes, sun-dried tomatoes offer more flavor than hothouse tomatoes or even canned tomatoes. This makes them ideal for adding to salads, topping a cheese pizza, adding pizzazz to a sauce. However, you can't use them as you would fresh or canned tomatoes, the flavor is different—more powerful. And the color is likely to bleed in slowly cooked soups or cream sauces, turning dishes that started out creamy white an unattractive beige. It makes the most sense to regard a sun-dried tomato as a unique ingredient with its own special flavor, just as a raisin bears only a passing resemblance to the grape from which it sprang.

What I've come to appreciate most about sun-dried tomatoes is that they have affinities for certain ingredients and combinations of ingredients. You cannot fail when you combine sun-dried tomatoes with cheese, garlic, or olive oil. Olives, seafood, and chicken harmonize well with sun-dried tomatoes. Likely seasonings include oregano, basil, thyme, and rosemary. Pizza and sun-dried tomatoes, rice and sun-dried tomatoes, sharp-tasting salad greens and sun-dried tomatoes . . . the combinations are endlessly rewarding.

The Tradition of Sun-Dried Tomatoes

That sun-dried tomatoes have an affinity for ingredients that are the mainstay of Italian cuisine should not be surprising, considering that sun-dried tomatoes marinated in oil were first prepared in southern Italy, where tomatoes and olive oil are both staples.

Tomatoes originated in the New World and were brought to Europe by the explorers in the sixteenth century. These first tomatoes were yellow, earning them the name of pomodoro, golden apples, in Italy. But it wasn't until two Jesuits returned from Mexico with seeds for a strain of red tomatoes that tomatoes gained a foothold in Italian cuisine, thriving in the warm, sunny climate.

In the south of Italy, where tomatoes are now a key ingredient in many dishes, the growing season lasts 120 days. To preserve tomatoes for use throughout the winter, tomatoes are dried in the sun. For this, Italians choose a small, firm tomato that is cultivated mainly for preserving; such tomatoes aren't eaten raw. The tomatoes grow in clusters, which are hung to dry in the sun. In some cases the tomatoes are then layered in glass jars with olive oil and herbs or chili peppers and sold as *pomodori secchi*. I have also heard of immigrant Italian women who mashed their sun-dried tomatoes into a paste with olive oil and stored the paste in crocks. The crocks were added to as tomatoes ripened and dried, and the paste was used all winter long to flavor sauces.

Sun-dried tomatoes packed in oil were introduced to the United States in 1980 by specialty food purveyors Dean and DeLuca. Their introduction coincided with the rising popularity of all foods Italian, and sun-dried tomatoes entered the American culinary scene.

Buying Sun-Dried Tomatoes

Sun-dried tomatoes are available from American producers as well as Italian importers. You can buy sun-dried tomatoes dry or packed in oil, or in paste form. Not all tomatoes are dried in the sun; some are dried in dehydrators. But since there is no appreciable flavor difference between sun-dried and dehydrator-dried tomatoes, I refer to them all as sun-dried tomatoes. Dry tomatoes, those not packed in oil, have the advantage of being less expensive and taking up less room in storage. If you buy them through the mail, shipping charges will be less. The disadvantage is that they must be refreshed in water before they can be used in a recipe.

The quality of sun-dried tomatoes varies from brand to brand, sometimes from season to season. Occasionally you will come across tomatoes that taste salty or seem tough. Most companies sell small packets of dried tomatoes, often in 3-ounce bags, so your investment isn't great if you don't like a particular brand.

If you should come across a brand that you find tough or salty, refresh the tomatoes in a few changes of hot water until the tomatoes taste good to you. You may need to soak tough tomatoes for up to 15 minutes or use them in recipes that contain a lot of liquid.

One advantage to buying the tomatoes packed in oil is that they are ready to be used. Again, there is some variation in flavor and tenderness, so if you can, taste before buying large quantities. You'll find tomatoes packed in oil in 3½-ounce and 7-ounce jars on the shelves of specialty food markets. In Italian delis and some specialty food stores you'll often find the tomatoes packed in huge glass jars, from which you can buy as much or as little as you need.

Sun-dried tomatoes packed in oil are pretty expensive ($15 to $20 a pound), in part because you are paying for the olive oil in which they are packed. This olive oil is worth it! The oil takes on a wonderful tomato flavor and can be used for dressing salads, as well as for adding flavor to sautés. Of course, you can refresh dried tomatoes in hot water, then drain and pack them in your own olive oil. This will give you the same supply of deliciously flavored oil, as well as a convenient supply of ready-to-use tomatoes—and it will save you money. In any event, since sun-dried tomatoes are mostly used as flavor accents, you will find that a pound goes a long way.

Some specialty food stores also carry sun-dried tomato paste in squeeze tubes and jars of dried tomato tapanade. Both are essentially the same product, which is a puree of sun-dried tomatoes, olive oil, garlic, and herbs. This puree is handy for adding flavor to soups, stews, and sauces, but rather expensive unless you make your own. Just mince a few garlic cloves in a food processor, add about a cup of sun-dried tomatoes marinated in oil, and process until smooth. You may want to add up to a tablespoon of olive oil to make a smooth consistency. Fresh herbs, such as oregano and basil, are optional. You can also make the paste by refreshing dried tomatoes in water for about 2 minutes, draining, and then pureeing them with olive oil. Keep the paste in a sealed jar in the refrigerator, where it will keep for several months. Use the puree by the heaping tablespoon; it adds a wonderful richness to a dish.

Drying Your Own Tomatoes

I've seen pictures of Italian women hanging tomatoes in clusters to dry slowly outside. Not many climates in this country allow for drying outside, but if you live in the Southwest or in Southern California, you might want to try it. Otherwise you can dry tomatoes in a food dehydrator or in a conventional oven.

How long it actually takes to dry tomatoes varies wildly, depending on the amount of rainfall during the growing season, the variety of tomato, the relative humidity of the air during the drying time, the efficiency of your dehydrator, and the color of your baking tins if you are drying in a conventional oven.

Regardless of how you plan to dry them, start with firm, ripe, Italian plum tomatoes. Cherry tomatoes, beefsteak tomatoes, and other good salad tomatoes are too watery for drying. Plum tomatoes, by contrast, have relatively few seeds and firm, meaty flesh.

It takes about 17 pounds of fresh tomatoes to make 1 pound of dried tomatoes. Or, to look at it another way, it takes about 8 pounds of tomatoes to fill a l-quart canning jar with dried tomatoes. I usually dry at least 30 pounds each summer (1 bushel), which leaves me with enough to cook with and give away half-pint jars as gifts.

After washing the tomatoes and allowing them to dry completely, remove the cores and slice the tomatoes in half, lengthwise. Some people prefer to partially slice open the tomatoes so they can be opened and closed like a book. These tomatoes can be packed more attractively in glass jars. It is probably faster and easier to slice through all the way. Lightly salting the tomatoes helps to draw out moisture, but some people prefer to skip this step.

If you are drying outside, arrange the tomatoes skin side down on screening and cover with netting to prevent insects from landing on the tomatoes. Leave outside to dry in the hot sun for about 48 hours. If the nights are cool, you may want to bring them inside.

If you have a food dehydrator, arrange the tomatoes skin side down on trays set about 2½ inches apart. Dry them on high for 10 to 16 hours. Begin checking after 10 hours and remove smaller tomatoes as they dry.

To dry in a conventional oven, arrange the tomatoes on wire racks set on baking tins. When the air is dry, you may be able to dry at 170°F., but watch them closely for burning. Prop open the oven door with a wooden spoon to increase air circulation. Every hour or so, rotate the orientation of the trays from front to back and rotate the position of the trays from the top to the bottom shelves. Toward the end of the drying time, you may want to turn the tomatoes over. Remove the tomatoes as they dry; in an oven the tomatoes are most likely to dry at differ-ent rates. Tomatoes on blackened baking tins will dry faster than those on shiny silver or gray pans.

When dry, the tomatoes will darken. They should feel leathery rather than brittle, but not at all tacky to the touch. Completely dry tomatoes will keep in plastic bags or airtight jars for years, although the color and flavor will deteriorate after a year. Don't store tomatoes like this unless they are completely dry; otherwise they will mold. To use, refresh for 2 to 10 minutes in boiling hot water, then drain and pack in glass jars. Cover with extra-virgin or cold-pressed pure olive oil. If you like, add herbs and garlic to the tomatoes. Stored in oil, the tomatoes will keep for at least a year.

If you will be using your tomatoes soon after drying, you may want to dry the tomatoes until they are leathery and then pack in olive oil, without refreshing them. The color of these tomatoes will be brighter red and the flavor will be excellent. You will have to experiment to get the texture right; if the tomatoes are too moist, they will disintegrate in the jar.

A very convenient way to handle dried tomatoes in quantities is to store it as paste, made as described earlier. Less olive oil is required so the cost of putting these tomatoes up is less. Also the paste can be frozen in small quantities and defrosted as needed.

Tomato Oil

The oil in which the tomatoes are packed takes on a deliciously fruity, tomato taste. This is true whether you buy commercially packed tomatoes or dry and pack your own. My favorite use for this oil is to drizzle it on fresh tomatoes and mozzarella.

You will have to experiment to see whether you prefer a green extra-virgin oil or a pure olive oil. As you use the oil in the jar, replenish with fresh oil to keep the tomatoes covered (very important for preventing molding).

About the Recipes in This Book

I have used mostly oil-packed tomatoes for this book. Because sun-dried tomatoes vary tremendously in size, I have used volume measures for almost all of the recipes. Even so there is room for judgment, depending on how tightly you pack the tomatoes into the measuring cup. But since the tomatoes are used for flavoring, a little more or a little less won't radically affect the outcome of the recipe. Indeed, if you find the recipes lacking in tomato punch, increase the amount called for; likewise, if the recipes strike you as overly pungent, decrease the tomato measure.

Almost all of the recipes call for either no salt at all or salt to taste. Again, judgment is required; your tomatoes may be more or less salty than mine.

The recipes are grouped traditionally—appetizers, salads, entrees, and so on. But that's not really how most people cook and eat these days. A few appetizers grouped together can make a light meal; many of the salads are intended to be served as main courses. And, of course, pasta and pizza both make satisfying meals by themselves, though they can be served as first courses.

Obviously, when one compiles recipes for a cookbook, one selects only the tried, true, and delicious. But, if someone were to ask me which recipes are the best to try first, I'd probably say the main course salads, which take so little time to prepare and are so satisfying, or maybe the pizzas, of which my family never tires. Or perhaps it's the pasta dishes that bring out the special taste of sun-dried tomatoes the best—it's so hard to say . . .

BRUSCHETTA

Bruschetta, grilled bread, has marked the start of many a meal in central Italy. Traditionally, it consists of thick slices of fresh Italian bread, grilled over charcoal, rubbed with garlic, brushed with olive oil, and sprinkled with salt and pepper. Traditionalists will claim the addition of sun-dried tomatoes and Parmesan cheese is gilding the lily. I say it's worth trying at least once. Use only bread that is home-baked or made in a high-quality Italian bakery. Steer clear of any bread that has been wrapped in plastic.

12 thick slices Italian bread
2 to 4 garlic cloves, crushed
Approximately 1/2 cup olive oil (from marinated tomatoes, if desired)
3/4 cup slivered sun-dried tomatoes marinated in oil
Coarsely ground black pepper to taste

Arrange the bread on a baking sheet and toast under the broiler until golden brown. Rub at once with the garlic. Brush with the olive oil. Sprinkle the tomatoes and pepper on top. Serve at once.

Yield: 6 servings

Bruschetta with Anchovies

To properly enjoy this, you must imagine yourself on the Mediterranean coast. With each bite, imagine that you had prepared a snack of bread, tomatoes, and anchovies and left it to bake on a rock in the sun. When you returned from your swim, the flavors of the yeasty bread, the fruity olive oil, the pungent tomatoes, and the salty anchovies had all melded together. The sun, the salty air—these are what this appetizer tastes like.

1/4 cup olive oil (from marinated tomatoes, if desired)
1/4 cup sun-dried tomatoes marinated in oil or 2 tablespoons sun-dried tomato puree (see page 8)
2 tablespoons anchovy paste
1 loaf Italian bread, sliced (halve or quarter the slices if they are large)
1/2 cup chopped fresh Italian parsley

Combine the olive oil, tomatoes, and anchovy paste in the food processor and puree to make a smooth paste.

Arrange the bread on a baking sheet and toast under the broiler until golden and crisp. Spread each piece with the tomato puree. Sprinkle with the parsley and serve.

Yield: 6 servings

Sun-Dried Tomato Finger Sandwiches

Sometimes the simplest recipes are the best. These dainty sandwiches can be served anywhere, any time—but I like them best served outside as a small bite to eat with a chilled glass of white wine. An ocean view and sea breezes add to the flavor.

Sun-dried tomatoes, marinated in oil
Baguette, thinly sliced
Cream cheese
Fresh whole basil leaves or tiny sprigs of watercress

Drain the tomatoes and cut into quarters. Spread slices of bread with cream cheese. Garnish each with a small, whole basil leaf or sprig of watercress and a piece of tomato and serve.

Yield: 6 servings

Cheesy Pita Triangles

1 cup cream cheese, at room temperature
1 cup unsalted butter, at room temperature
1 cup freshly grated Parmesan cheese
1/2 cup sun-dried tomatoes marinated in oil
2 tablespoons oil from sun-dried tomatoes
Basil leaves
6 pita pockets

Combine the cream cheese, butter, and Parmesan cheese in a food processor and beat until very smooth. Cut 4 of the tomatoes into thin strips; set aside. Whirl the remaining tomatoes, oil, and 1/2 cup of the cream cheese mixture in a blender until all are smoothly pureed. Combine the tomato puree with the cream cheese mixture and beat to blend. Cover and chill for about 20 minutes or until firm. Preheat the oven to 400°F.

To prepare the pita toast triangles, split each pita in half to make 2 rounds. Cut each round into 6 triangles. Place in a single layer on cookie sheets. Bake in the preheated oven for 3 minutes or until lightly toasted. Let cool.

Spread the cheese on the toast and top each with a basil leaf or a tomato strip. Arrange on a large platter and serve.

Yield: 8 to 10 servings

Striped Tomato Brie

The rich tastes of brie and walnuts find the perfect complement in the tomatoes. This makes a lovely centerpiece on a buffet table.

1 (8-ounce) round ripe brie
1/4 cup butter, softened
1/3 cup chopped sun-dried tomatoes marinated in oil
2 tablespoons chopped fresh basil or 2 teaspoons dried
1 tablespoon minced shallots
1/4 cup coarsely chopped toasted walnuts
1/4 cup finely chopped toasted walnuts

With the foil wrapping in place, chill the cheese in the freezer until firm but not frozen, about 1 hour.

With a sharp knife, halve the cheese horizontally; set aside.

Beat the butter until creamy. Then beat in the tomatoes, basil, and shallots. Mix in the 1/4 cup coarsely chopped walnuts. Spread the tomato mixture evenly on the cut side of one cheese half. Cover with the remaining cheese, cut side down; press gently. Roll the cheese on its side in the 1/4 cup finely chopped toasted walnuts to coat the tomato layer. Cover and chill until firm. Serve cut into thin wedges. Accompany with crackers and baguette slices.

Note: The completely assembled brie can be wrapped and frozen for up to 1 week.

Yield: 8 to 10 servings

Brie Pesto Torte

Here's another elegant brie dish that makes a lovely addition to a buffet table. It should be prepared a day ahead to allow the flavors to blend and mellow. Pesto can be made in quantities when basil is in season and frozen for use all year-round.

1 cup fresh basil leaves
2 garlic cloves
3/4 cup freshly grated Parmesan cheese
1/4 cup pine nuts
3/4 cup olive oil
Salt and pepper to taste
2-pound wheel brie, well chilled
1/2 cup slivered sun-dried tomatoes marinated in oil
1/2 cup toasted pine nuts
Fresh basil leaves

Make the pesto by combining the basil, garlic, Parmesan, and 1/4 cup pine nuts in a food processor and pureeing until smooth. With the motor running, slowly add the olive oil until you have the desired consistency. Add salt and pepper to taste.

With a sharp knife, slice the brie in half horizontally. Spread the pesto on the bottom half of the brie. Sprinkle with the tomatoes. Top with the second half of the cheese, cut side down. Chill overnight.

Bring to room temperature and decorate with the toasted pine nuts and fresh basil leaves just before serving.

Yield: 10 to 20 servings

Grilled Stuffed Grape Leaves

You can use grape leaves that you collect yourself and blanch in boiling water until pliable, but I find it is easier to use the commercial leaves. With home-grown leaves, you may need to tie the stuffed leaves with cotton thread to prevent them from spilling the cheese and tomatoes onto the grill. This is a great appetizer to make when you are already firing up the grill for a barbecue.

16 grape leaves
1/2 cup sun-dried tomatoes marinated in oil, well drained
4 ounces creamy goat cheese, cut in 1/4-inch slices

Prepare a fire in the grill.

Carefully remove the grape leaves from the jar, rinse, and pat dry. Or blanch in boiling water until flexible, drain, and pat dry. Lay the leaves out on a flat work surface. Place a tomato in the center of each grape leaf and lay a slice of goat cheese over the tomato. Bring the leaf up around the tomato and cheese, folding so there is no cheese or tomato exposed. Continue until all the leaves have been filled.

Place the leaves on the grill over hot coals and grill for 2 to 3 minutes on each side. Serve immediately.

Yield: 4 servings

Marinated Stuffed Grape Leaves

1/2 pound goat cheese, at room temperature
1 (3-ounce) package cream cheese, at room temperature
1 cup chopped walnuts
1/2 cup slivered sun-dried tomatoes marinated in oil, well drained
24 grape leaves
1 cup olive oil
1/4 cup red or white wine vinegar
Salt and pepper to taste

Combine the goat cheese and cream cheese in a food processor and process until smooth. Fold in the nuts and tomatoes.

Carefully remove the grape leaves from the jar, rinse, and pat dry. Or blanch the leaves in boiling water, drain, and pat dry. Lay the leaves out on a flat work surface. Place 2 to 3 tablespoons of the cheese mixture on the center of each leaf. Fold the sides in and over the mixture, then roll up so no filling is exposed. Place seam side down in a glass baking dish.

Combine the oil, vinegar, and salt and pepper. Pour over the grape leaves. Cover and refrigerate overnight. Serve cold or at room temperature.

Yield: 6 servings

Baked Stuffed Clams

Salty, sweet sun-dried tomatoes pair well with salty, sweet clams. These clams can be prepared early in the day, then refrigerated and baked just before serving.

12 medium-size cherrystone or littleneck clams, well scrubbed
1/4 cup dry white wine
1 tablespoon olive oil from marinated tomatoes
2 shallots, minced
2 garlic cloves, minced
1/4 cup minced celery
2/3 cup dry bread crumbs
1 teaspoon dried thyme
1/4 cup chopped fresh parsley
1/4 cup slivered sun-dried tomatoes marinated in oil
Black pepper to taste
2 tablespoons dry bread crumbs
Approximately 2 tablespoons olive oil from marinated tomatoes

Combine the clams with the wine in a large saucepan. Cover and steam until the clams just open, 5 to 10 minutes. Reserve the broth. Remove the clams from the shells, chop, and set aside. Wash the shell halves and set in a shallow baking pan.

Preheat the oven to 350°F.

Heat the 1 tablespoon oil in a small sauté pan. Add the shallots, garlic, and celery; sauté until slightly limp, about 2 minutes. Combine with the clams, 2/3 cup bread crumbs, thyme, parsley, tomatoes, and black pepper in a small mixing bowl. Stir in enough reserved broth to moisten the mixture.

Distribute the clam and bread crumb mixture evenly among the shells. Sprinkle with the remaining 2 tablespoons bread crumbs and drizzle the remaining oil over each.

Bake in the preheated oven for about 10 minutes (a few minutes longer if the clams are chilled). Place under the broiler and broil until brown, about 4 minutes. Serve warm.

Yield: 4 servings

Stuffed Mushrooms

Stuffed mushrooms are a favorite, whether they are served as appetizers or side dishes to a roast. The mushrooms can be stuffed early in the day and refrigerated until serving time.

16 large fresh mushrooms
4 tablespoons olive oil
6 tablespoons finely minced onion
1/2 cup minced sun-dried tomatoes marinated in oil
4 tablespoons dry white wine
1/2 cup grated cheeses (Parmesan, Swiss, etc.)
3 tablespoons dry bread crumbs
1/2 cup minced fresh parsley
1/2 teaspoon dry tarragon
2 tablespoons cream (optional)
Salt and pepper to taste
4 tablespoons butter

Preheat the oven to 350°F. Remove the stems from the mushrooms. Finely mince the stems and set aside. Brush the mushroom caps with 1 tablespoon of the oil. Place the caps, cavity side up, in a baking pan.

Heat the remaining 3 tablespoons oil in a large frying pan. Add the onion and sauté until soft, 2 to 3 minutes. Add the minced mushroom stems and continue to sauté until lightly browned, about 4 minutes. Add the minced tomatoes and wine; boil rapidly until the liquid is almost evaporated. Remove from the heat.

Set aside a few tablespoons of the cheese for the topping. Mix in the remaining cheese, along with the bread crumbs and herbs. If the mixture is too dry to hold together, add up to 2 tablespoons cream to moisten. Season with salt and pepper.

Fill the mushroom caps with heaping spoonfuls of stuffing. Top each with a dot of butter and a sprinkle of grated cheese. Bake for 20 minutes. Serve hot.

Yield: 4 servings

Chevre Cheesecake

6 ounces sesame breadsticks or crackers
1/2 cup butter, melted
2 pounds goat cheese, at room temperature (rind removed)
1 pound cream cheese, at room temperature
3 eggs
1 teaspoon crushed rosemary
Salt and pepper to taste
1/2 cup minced sun-dried tomatoes marinated in oil

Generously butter a 9-inch springform pan. Preheat the oven to 325°F.

Process the breadsticks in a food processor until you have fine crumbs. Combine with the melted butter. Reserve 2 tablespoons of the crumb mixture and press the remaining crumbs along the bottom and up the sides of the pan. Chill.

Combine the goat cheese, cream cheese, eggs, rosemary, and salt and pepper in a food processor. Process until well mixed. Spoon into the crust and sprinkle the reserved crumbs on top.

Set the cheesecake in a larger pan and fill with water halfway up the sides of the pan. Bake for about 45 minutes, until puffed and lightly browned. Cool and refrigerate overnight. Garnish with the minced sun-dried tomatoes and serve with crackers.

Yield: 10 to 12 servings

MEDITERRANEAN ZUCCHINI BITES

2 teaspoons dried oregano
4 anchovy fillets
2 tablespoons drained capers
1 tablespoon lemon juice
1 large garlic clove
1/4 pound mushrooms
1/3 cup pitted ripe olives
1/2 cup sun-dried tomatoes marinated in oil
Black pepper to taste
Cayenne pepper to taste
2 pounds zucchini (about 1 1/3 inches in diameter), ends trimmed

Preheat the oven to 425°F.

In a food processor fitted with a steel blade, pulse the oregano, anchovies, capers, lemon juice, and garlic until coarsely chopped. Add the mushrooms and olives; process until finely minced. Mix in the tomatoes, black pepper, and cayenne. Set aside.

Cut the zucchini into l-inch rounds. With a melon bailer, scoop out a cavity from one of the cut sides of each piece; do not cut through to the other side. (Reserve the scooped-out bails for another use.) Stuff each cavity with a generous 2 teaspoons of the tomato mixture. Place the zucchini pieces on an ungreased baking sheet. Bake in the preheated oven for 5 minutes or until hot. Serve hot.

Yield: 2 1/2 dozen appetizers

Devilish Eggs

The classic deviled egg is flavored with mayonnaise and mustard. This tomato version provides a nice change.

6 hard-cooked eggs, peeled
1/2 cup mayonnaise
4 tablespoons sour cream
1 tablespoon vinegar
5 tablespoons minced sun-dried tomatoes marinated in oil

Slice the eggs in half lengthwise, remove the yolks, and set aside the whites.

Crumble the yolks. Combine with the mayonnaise, sour cream, vinegar, and 4 tablespoons of the dried tomatoes. Spoon the mixture back into the empty egg whites or use a pastry bag to pipe in the filling.

Sprinkle the remaining 1 tablespoon of dried tomatoes on top for color. Serve immediately or chill for a few hours.

Yield: 12 deviled eggs

Grilled Swordfish Salad

On a hot summer night, nothing is more satisfying than a warm seafood salad. In this recipe, the swordfish is marinated in a red wine vinaigrette, then grilled and placed on a bed of greens. To round out the salad, boiled new potatoes and green beans are added. If the potatoes and beans come from the garden, so much the better.

1 cup olive oil
10 tablespoons red wine vinegar
4 large garlic cloves, minced
4 teaspoons minced fresh tarragon or 2 teaspoons dried
2 tablespoons minced fresh basil or 2 teaspoons dried
Salt and pepper to taste
4 swordfish steaks, about 1/2 pound each
1 1/2 pounds new potatoes, halved or quartered
1 1/2 cups green beans cut in 1-inch pieces
1/2 cup diced scallions
8 to 10 cups mixed salad greens, including some red-leaf lettuce or radicchio
3/4 cup slivered sun-dried tomatoes marinated in oil

About 1 hour before serving, make a marinade for the fish by mixing together the oil, vinegar, garlic, and herbs. Season with salt and pepper. Slice the steaks into large chunks, about 6 per steak. Toss the fish with about half of the marinade, cover, and refrigerate until about

30 minutes before you are ready to serve. Set aside the remaining marinade.

Boil the potatoes in their jackets until tender, 8 to 10 minutes. Drain and toss with the reserved marinade. Blanch the beans in boiling water until slightly tender and bright green, about 2 minutes. Drain and rinse the beans in cold water to stop the cooking. Add to the potatoes, along with the scallions.

Arrange the greens on individual dinner plates.

Thread the fish onto presoaked bamboo skewers. Grill over a hot fire for about 10 minutes, turning once and basting often with the marinade. Grill until the fish is firm to the touch and white.

Divide the potato and green bean mixture among the plates and arrange on top of the greens. Place the fish on top of the vegetables. Scatter the sun-dried tomatoes over all. Drizzle the remaining marinade on top, if desired. Serve at once.

Grilled Tuna Salad

Substitute fresh tuna steaks for the swordfish and proceed with the recipe as above.

Warm Beef Salad

Caper Vinaigrette

1 tablespoon red wine vinegar
1½ tablespoons lemon juice
1 teaspoon minced capers
⅓ cup olive oil
Freshly ground black pepper to taste

Salad

9 to 10 cups mixed salad greens
1 to 2 small zucchini
¾ cup slivered sun-dried tomatoes marinated in oil
1 to 2 carrots
1 pound London broil
1 red onion, sliced in rings

To make the vinaigrette, combine the vinegar, lemon juice, and capers in a small bowl. Whisk in the olive oil. Season with pepper. Set aside. Whisk again before serving.

Wash and dry the salad greens. Arrange on dinner plates. Slice the zucchini into thin rounds. Blanch in boiling water for about 1 minute, then plunge into cold water to stop the cooking process. Dry and add to the greens along with the tomatoes. Toss very gently to mix the vegetables with the greens. Cut the carrots into shoestrings or matchsticks. Add to the salad, arranging the carrots around the perimeter of the plates.

Grill or broil the meat until rare. This will take about 10 minutes, depending on how hot the fire is. Thinly slice the meat against the grain. Lay 4 to 5 slices of the meat over the top of each salad. Drizzle salad dressing over all. Garnish with the red onion rings. Pass extra salad dressing at the table.

Yield: 4 servings

Grilled Turkey Salad

Turkey

2 shallots
3 garlic cloves
1 tablespoon sun-dried tomato puree (page ?)
1/3 cup olive oil
1/3 cup lime juice
1 tablespoon balsamic vinegar
1/4 teaspoon salt
1 teaspoon cumin
1 teaspoon chili powder
3 to 4 turkey thighs (about 2 pounds)

Salad

8 to 10 cups mixed salad greens
10 scallions, sliced in 1-inch pieces
1 red bell pepper, sliced in rings
2 avocados, sliced
1 cup slivered sun-dried tomatoes in oil
1/4 cup olive oil
2 tablespoons balsamic vinegar
2 tablespoons lime juice

Combine the shallots, garlic, and tomato puree in a food processor and process until smooth. Add the olive oil, lime juice, vinegar, salt, cumin, and chili powder and mix well. Slice the turkey into 2-inch pieces. Place the turkey pieces in a glass dish. Pour the marinade over, and turn to coat the turkey. Set aside to marinate for about 3 hours in the refrigerator or for 1 hour at room temperature. Prepare the fire in the grill.

Arrange the greens on dinner plates. On top of the greens, arrange the scallions, red pepper, avocados, and tomatoes. In a small bowl, combine the oil, vinegar, and lime juice to make a dressing. Drizzle the dressing over the salads. Grill the turkey over white-hot coals for about 20 minutes, turning once after 10 minutes. Arrange on top of the salads, and serve.

Yield: 4 servings

Grilled Chicken Salad

Appreciated for its hearty, peppery flavor, arugula has been popular for many years among home gardeners under the name rocket.

Dressing

2 tablespoons plain yogurt
2 tablespoons mayonnaise
2 tablespoons white wine vinegar
2 tablespoons olive oil
2 tablespoons white wine
2 to 3 garlic cloves, minced
1 tablespoon minced fresh parsley
Freshly ground black pepper to taste

Salad

4 boneless, skinned chicken breasts
8 teaspoons olive oil
8 to 10 cups mixed salad greens
1 (6-ounce) jar marinated artichoke hearts
2 cups slivered sun-dried tomatoes marinated in oil

First prepare the dressing by combining all the ingredients and mixing well. Set aside for at least 30 minutes to allow the flavors to blend.

Brush each chicken breast with about 1 teaspoon oil. Place the breasts oiled side down on a grill over white hot coals and grill for about 15 minutes, turning once halfway through the cooking time. Before turning, brush the breast with the remaining oil.

While the chicken cooks, arrange a bed of greens on 4 dinner plates. Arrange the artichoke hearts on the greens.

Remove the chicken from the grill when done, and slice each piece about 1/2 inch thick. Arrange the chicken pieces over the greens. Sprinkle the tomatoes over all. Spoon a little dressing over each salad, and pass the remaining dressing at the table.

Yield: 4 servings

Tomato and Chicken Salad

This recipe is a wonderful example of the transforming power of sun-dried tomatoes. With the tomatoes, an ordinary dish like chicken salad becomes something special. I like to serve this on a bed of greens, but it also makes an excellent sandwich filling.

2 cooked chicken breasts, diced (about 4 cups)
3 cups diced celery
2 cups diced green or red bell pepper
1 cup chopped walnuts
1/2 cup chopped fresh chives or 3 tablespoons dried
1 cup mayonnaise
1/4 cup lime juice
Salt and pepper to taste
1 cup chopped sun-dried tomatoes marinated in oil

In a large mixing bowl, combine the chicken, celery, pepper, walnuts, and chives. Add the mayonnaise and lime juice and mix well. Season with salt and pepper (be generous with the pepper). Lightly mix in the tomatoes. To allow the flavors to blend, let the salad stand for 15 to 30 minutes before serving.

Yield: 6 servings

Warm Shrimp and Avocado Salad

The sweetness of the shrimp and the tomatoes combines wonderfully with the creamy avocado. Serve it on a bed of sharp-tasting greens for the full effect.

8 cups mixed salad greens, including arugula, radicchio, and other sharp-tasting greens
2 to 4 avocados, peeled and sliced
2 cups slivered sun-dried tomatoes marinated in oil
1/2 cup olive oil
2 pounds shrimp, peeled and deveined
1 teaspoon dried tarragon, basil, or oregano
1/4 cup balsamic vinegar
Freshly ground black pepper to taste

Arrange the greens on 4 dinner plates. Arrange the avocado slices on the greens. Sprinkle each plate with 1/2 cup tomatoes.

In a large skillet, heat the oil. Add the shrimp and herbs. Sauté until the shrimp are pink and firm, 4 to 5 minutes. Remove the shrimp from the pan with a slotted spoon and arrange on the salads. Add the vinegar to the pan and cook for about 2 minutes, stirring constantly. Drizzle the pan juices over the salads. Sprinkle with black pepper and serve at once.

Yield: 4 servings

Sonoma Caesar Salad

Here sweet, chewy, sun-dried tomatoes, the newest addition to the classic caesar salad, provide a counterpoint to the salty anchovies.

Garlic Croutons

3 tablespoons sun-dried tomato oil
2 garlic cloves, chopped
3 cups bread cubes

Dressing

2 garlic cloves
1 (2-ounce) can anchovies, drained
1/2 cup olive oil
1/2 cup lemon juice
1/2 teaspoon Dijon-style mustard
1/2 teaspoon Worcestershire sauce
Coarsely ground black pepper to taste

Salad

1 head romaine lettuce, washed and dried
1/2 cup sun-dried tomatoes in oil
1/2 cup grated Parmesan cheese

To prepare the croutons preheat the oven to 325°F. Warm the 3 tablespoons of oil in a skillet with the garlic. Add the bread cubes and toss to coat with oil. Transfer to a baking sheet, spreading out the cubes to a single layer. Bake in the preheated oven, tossing occasionally, until crisp and golden, 12 to 15 minutes.

To prepare the dressing, combine the remaining 2 cloves garlic and anchovies in a food processor or blender and pulse until finely minced. Add the remaining 1/2 cup oil, lemon juice, mustard, and Worcestershire, and blend thoroughly. Season with black pepper.

Gently tear the lettuce into a salad bowl. Top with the croutons, tomatoes, and cheese. Drizzle the dressing over. Toss and serve immediately.

Yield: 4 to 6 servings

PESTO PASTA SALAD

This salad has all the tastes of summer—pesto, tomatoes, crisp vegetables. Best of all, it can be made easily in winter with frozen pesto and sun-dried tomatoes.

To make pesto, combine about 2 cups of basil leaves with some garlic, about 1/2 cup of Parmesan cheese, and 1/4 cup of pine nuts in a food processor or blender. With the machine running, add olive oil until you have a smooth paste, about 1/2 cup. Season to taste with salt and pepper. Store in the refrigerator in an airtight container with a layer of olive oil on top to seal out any air. Or freeze it. You can buy prepared pesto in many specialty food stores.

12 ounces uncooked fusilli, twist, or seashell pasta
1/4 cup olive oil
1/4 cup red wine vinegar
Juice of 1 lemon
1/3 to 1/2 cup pesto
1/2 cup slivered sun-dried tomatoes marinated in oil
1 cup fresh or frozen peas
1 carrot, grated
1/2 cup toasted slivered almonds
Salt and pepper to taste

Cook the pasta in plenty of boiling salted water according to the package directions. Rinse in cold water, then drain.

In a large bowl, combine the oil, vinegar, lemon juice, and pesto. Add the pasta and toss to mix. Add the remaining ingredients and toss to mix. Season with salt and pepper.

Yield: 4 to 6 servings

CÔTE D'AZUR FUSILLI SALAD

This salad combines ingredients from the south of France—tomatoes, olives, goat cheese, and basil.

1/2 pound dry egg and spinach fusilli
1 tablespoon olive oil
2 garlic cloves, minced
2 teaspoons chopped fresh basil
8 sun-dried tomatoes marinated in oil, coarsely chopped
8 to 10 Kalamata olives, pitted and quartered (or any black Greek olive)
2 tablespoons pine nuts
Balsamic vinegar
1/4 pound goat cheese, cut into 1/4-inch chunks
2 tablespoons chopped fresh chives

Cook the pasta in plenty of boiling water according to the manufacturer's directions. Rinse in cold water and drain. Transfer to a serving bowl or platter and toss with the olive oil.

Mix the garlic and basil with the pasta. Then fold in the tomatoes, olives, and pine nuts. Sprinkle the salad with balsamic vinegar to taste, about 1 to 2 tablespoons. Add the cheese last and toss lightly. Garnish the salad with the chives and serve.

Yield: 4 to 6 servings

Antipasto Pasta Salad

Italian Dressing

1 1/2 tablespoons red wine vinegar
1 1/2 tablespoons white wine vinegar
1 garlic clove, minced
1/2 teaspoon minced capers
Approximately 3 tablespoons olive oil
Freshly ground black pepper to taste

Pasta Salad

1/2 pound uncooked spaghetti rings
1 tablespoon olive oil
6 to 8 sun-dried tomatoes marinated in oil
6 canned artichoke hearts, drained
1/4 pound dry salami, sliced 1/4 inch thick
1/4 pound fontina cheese
8 spicy green Italian olives
4 mild marinated cherry peppers
4 hot marinated cherry peppers
4 hot marinated peperoncini peppers

To make the dressing, whisk together the red and white wine vinegars with the garlic and capers in a small bowl. Slowly add the oil, continuing to whisk. Add enough oil to achieve a good acid balance. Grind the black pepper into the dressing.

Cook the pasta in plenty of boiling water according to the manufacturer's directions. Rinse in cold water and drain. Transfer to a serving bowl and toss with the olive oil.

Quarter the tomatoes, the artichoke hearts, and the salami slices. Cut the cheese into bite-size chunks. Halve and pit the olives, reserving a few whole ones for the garnish. Drain and slice the peppers into thin ringlets, reserving a few whole ones for the garnish.

Combine all the ingredients with the pasta and toss with the dressing. Garnish the salad with the whole olives and peppers and serve.

Yield: 6 to 8 servings

Tuna Pasta Salad

It's quite possible you have all the necessary ingredients for this salad in your cupboard and refrigerator.

1/2 pound uncooked wagon wheel or bowtie pasta
1 tablespoon olive oil
1 (8-ounce) can water-packed albacore tuna
1 to 2 tablespoons capers
6 to 8 sun-dried tomatoes marinated in oil
2 tablespoons chopped fresh parsley
1/2 to 3/4 cup fresh or frozen peas
Parsley sprigs

Cook the pasta in plenty of boiling water according to the manufacturer's directions. Rinse in cold water and drain. Transfer to a serving bowl and toss with the olive oil.

Drain the tuna and chop coarsely. Mince the capers if they are large; keep them whole if they are small. Chop the tomatoes into bite-size pieces.

Toss the pasta with parsley. Add the capers, tomatoes and peas, and toss again. Add the tuna last, and mix it just enough to distribute it. Garnish the salad with parsley sprigs and serve.

Yield: 4 to 6 servings

Potato Salad

With no mayonnaise, this colorful salad is perfect for picnics.

6 cups peeled and thinly sliced potatoes (6 to 8 potatoes)
1/2 cup olive oil (some can come from the marinated tomatoes)
1/4 cup white wine vinegar
1/2 cup sliced scallions, including green tops
1 red pepper, diced
1 green pepper, diced
1 cup chopped fresh parsley
2/3 cup slivered sun-dried tomatoes marinated in oil
Salt and freshly ground black pepper to taste

Boil the potatoes in salted water to cover until just tender, about 5 minutes. Drain.

In a large salad bowl, combine the potatoes with the oil and vinegar. Let cool at room temperature. The potatoes will absorb most of the dressing.

When the potatoes are cool, add the remaining ingredients, seasoning with salt and plenty of pepper. Let stand for at least 30 minutes to allow the flavors to blend.

Yield: 6 servings

MEDITERRANEAN TURKEY RICE SALAD

This salad makes a wonderful casual supper. Take it along for a picnic, or serve it at home with a cooked vegetable, such as steamed asparagus. Breadsticks finish the meal.

1 pound turkey breast or thigh cutlets
1/2 cup plain yogurt
2 garlic cloves, minced
1 small onion, finely chopped
3 cups cooked brown or white rice
1 1/2 cups cooked wild rice
1/2 cup chopped sun-dried tomatoes marinated in oil
1 cucumber, peeled, seeded, and finely chopped
1 cup chopped fresh parsley
1/2 cup chopped scallions, onion, or chives
1 tablespoon dried mint
1/2 cup olive oil
1/4 cup balsamic vinegar
Salt and pepper to taste

A few hours before you are ready to prepare the salad, combine the turkey with the yogurt, garlic, and onion. Mix well to coat. Cover and refrigerate.

Heat a nonstick or lightly oiled skillet over medium-high heat. Add the turkey and cook until lightly browned, about 12 minutes, turning the turkey halfway through the cooking time.

Remove the turkey from the skillet and dice. In a large bowl, combine the turkey with the rice, wild rice, tomatoes, cucumber, parsley, scallions, and mint. Pour the oil and vinegar over and toss. Add salt and pepper. Allow the salad to sit for at least 30 minutes before serving to allow the flavors to develop.

Yield: 4 servings

Goat Cheese and Tomato Salad

The relative lack of popularity of goat cheese is a good thing. No one has yet come up with a marketing scheme to package it in aerosol cans or individual plastic slices, and no manufacturer has denatured its exquisite, goaty essence. You'll find a few specific varieties of goat cheese in large specialty stores, but most goat cheeses are farmhouse products simply labeled as "chevre." Any variety—pungent or mild—will do here.

2 cups crumbled or diced goat cheese
2 tablespoons drained capers
3 sun-dried tomatoes marinated in oil, slivered
2 tablespoons white wine vinegar
2 teaspoons prepared mustard
1/4 cup olive oil (from the marinated tomatoes)
Salt and freshly ground black pepper to taste
2 heads Boston lettuce

Combine the goat cheese in a small bowl with the capers and tomatoes.

In another bowl, whisk together the vinegar and prepared mustard. Gradually add the oil, beating briskly with the whisk. Season with salt and pepper. Pour the dressing over the cheese and tomatoes.

Arrange the lettuce on 6 to 8 salad plates. Spoon the cheese mixture over the lettuce and serve.

Yield: 6 to 8 servings

Cucumber Yogurt Salad

Salting the cucumbers draws out excess water from the flesh and makes it extra crisp, so don't skip this step. If you prefer, rinse the salt from the cucumbers under running tap water, then season to taste with salt and pepper. If you can, use cucumbers that have not been waxed and don't peel.

8 cups sliced cucumbers
1/2 teaspoon salt
1 to 2 tablespoons chopped fresh basil or 1 to 2 teaspoons dried
1/2 cup chopped walnuts
1/2 cup plain yogurt
2 sun-dried tomatoes marinated in oil, slivered
Freshly ground black pepper to taste

Combine the cucumbers and salt in a colander and toss to mix. Place the colander inside a bowl to catch excess water. Weight the cucumbers with a heavy object (such as a canning jar filled with water) and set aside for 30 minutes.

In a large bowl, combine the cucumbers, basil, and walnuts. Toss gently. Add the yogurt and toss gently to mix. Fold in the tomatoes. Season to taste with pepper. Serve at once.

Yield: 4 servings

Tomato and Corn Salad

This finely chopped salad can be served on a bed of greens, or offered as a side dish or relish. It makes a nice companion to burgers.

4 cups fresh (about 6 ears) or frozen corn kernels
1 cup finely diced red pepper
2/3 cup finely diced red onion
1/2 cup slivered sun-dried tomatoes marinated in oil
1/2 cup chopped fresh parsley
2 tablespoons mayonnaise
2 tablespoons plain yogurt
2 tablespoons white wine vinegar
2 tablespoons olive oil (from marinated tomatoes)
Salt and pepper to taste

If you are using fresh corn, strip the kernels from the ears and briefly blanch in boiling water, about 1 minute. Rinse in cold water to stop the cooking. If you are using frozen corn, defrost and drain.

Combine the corn, red pepper, onion, tomatoes, and parsley in a medium-size salad bowl. In a small bowl, combine the mayonnaise, yogurt, vinegar, and oil. Pour over the salad and toss to coat. Season with salt and pepper. If possible, let stand for at least a few minutes to allow the flavors to blend. This salad can be held for several hours in the refrigerator before serving.

Yield: 4 to 6 servings

Basic Pizza Dough

The recipe can be doubled without any problem (double everything except the yeast). You can freeze the dough after it has risen. Just wrap the ball of dough in plastic wrap and use it within 3 months. Defrost the frozen dough completely before using it. The dough will not have as much elasticity as fresh dough, so be careful when stretching it to avoid tears.

1 tablespoon (1 package) active dry yeast
2/3 cup warm water (105°F to 115°F)
1 tablespoon olive oil
1 teaspoon salt
Approximately 1 1/2 cups all-purpose flour

Sprinkle the yeast over the water in a large bowl. Stir to dissolve and let stand for about 5 minutes. Then mix in the remaining ingredients, reserving about 1/2 cup of the flour. Turn the dough out onto a lightly floured board and knead in as much of the remaining flour as needed to make a soft, elastic dough. Place the dough in a lightly oiled bowl, turning to coat the entire surface of the dough with the oil. Cover and let rise in a warm place for 45 to 60 minutes.

When the dough has doubled in bulk, punch it down. Stretch the dough, without tearing it, to fit a 10-inch pizza pan or cookie sheet. If you are a novice at stretching pizza dough, you may prefer to roll it on a lightly floured surface. Carefully transfer the dough to the baking sheet and finish stretching it to its final shape with a 1/2-inch rim around the edge.

Use with the topping of your choice. Bake at 500°F for 10 to 15 minutes.

Yield: 10-inch round pizza crust

Double-Cheese Sausage Pizza

I make pizza about once a week—when the energy for something more ambitious is lacking. I serve it with a green salad and everyone in the family is happy. This is an especially easy pizza to make, requiring little in the way of planning and forethought.

Basic Pizza Dough (page 41)
1/4 pound hot or sweet Italian sausage, sliced into 1/2-inch pieces
1/2 pound provolone cheese, sliced
3/4 cup slivered sun-dried tomatoes marinated in oil
1/2 pound mozzarella cheese, grated

Prepare the pizza dough according to the recipe directions. Stretch to fit a lightly oiled 10-inch pizza pan or baking sheet, pushing up a 1/2-inch edge all around.

In a lightly oiled or nonstick skillet, brown the sausage, stirring frequently, about 8 minutes. Remove from the skillet with a slotted spoon and drain on paper towels.

Leaving a 1/2-inch border all around, cover the surface of the pizza with the provolone. Sprinkle the tomatoes and sausage on next. Cover with the grated cheese.

Bake in a preheated 500°F. oven for about 10 minutes, until the cheese begins to brown. Allow to stand for about 5 minutes before slicing and serving.

Yield: 4 servings

Leek and Feta Cheese Pizza

The combination of leeks, feta cheese, and sun-dried tomatoes makes a tasty pizza. For a special touch, add a half pound of shrimp.

Basic Pizza Dough (page 41)
1/4 cup butter
4 cups sliced leeks (about 1 pound)
1 to 2 tablespoons olive oil (optional)
1/2 pound shrimp, peeled and deveined (optional)
1 cup slivered sun-dried tomatoes marinated in oil
Freshly ground black pepper to taste
1/2 pound feta cheese, crumbled

Prepare the pizza dough according to the recipe directions. Stretch to fit a lightly oiled 10-inch pizza pan or baking sheet, pushing up a 1/2-inch rim all the way around. Preheat the oven to 450°F.

Melt the butter in a large skillet. Add the leeks and sauté until limp, about 3 minutes. Sprinkle on top of the dough.

If you are using the shrimp, add the oil to the skillet and heat over medium heat. Add the shrimp and sauté until pink and firm, 2 to 3 minutes. Sprinkle on top of the leeks. Sprinkle the tomatoes over, then the black pepper, then the cheese.

Bake for 15 to 20 minutes, until the crust is golden and hard, and the cheese is slightly browned. Serve hot.

Yield: 4 servings

ORB WEAVER PIZZA

This recipe originated at Orb Weaver Farm, makers of Orb Weaver cheese, a wonderful mild cheese.

Basic Pizza Dough (page 41)
2 tablespoons olive oil
2 cups thinly sliced onions
1 green pepper, chopped
1 cup chopped sun-dried tomatoes marinated in oil
1 cup chopped black olives (optional)
8 ounces Orb Weaver Farmhouse Cheese or Monterey jack, grated

Prepare the pizza dough according to the recipe directions. While the dough rises, prepare the topping.

Heat the olive oil in a skillet. Add the onions and sauté slowly over low heat until they are golden, about 5 minutes.

Preheat the oven to 500°F. Lightly oil a large pizza pan or baking sheet. Fit the dough onto the pizza pan, pushing up a 1/2-inch edge all the way around.

Spread the onions on the dough. Sprinkle the green pepper, tomatoes, and olives on top. Cover with the grated cheese.

Bake for 10 to 15 minutes, until the crust is golden and the cheese is melted and bubbly. Let stand for about 5 minutes before slicing and serving.

Yield: 4 servings

Ricotta Artichoke Pizza

The ricotta makes a creamy pizza topping. Its bland flavor complements the pungent tomatoes nicely.

Basic Pizza Dough (page 41)
4 garlic cloves
12 basil leaves
1 1/2 pounds ricotta cheese
Olive oil
1/2 cup slivered sun-dried tomatoes marinated in oil
6 tablespoons freshly grated Parmesan cheese
1/2 cup diced green pepper
1 (14-ounce) can artichoke hearts, drained and sliced

Prepare the pizza dough according to the recipe directions. Stretch to fit a lightly oiled 10-inch pizza pan or baking sheet, pushing up a 1/2-inch edge all the way around. Preheat the oven to 500°F.

Mince the garlic and basil in a food processor. Add the ricotta and beat until creamy.

Spread a little oil over the dough. Spread the ricotta on top. Sprinkle the tomatoes, then the Parmesan cheese on top of the ricotta. Then sprinkle the green pepper and artichoke hearts on top.

Bake for about 15 minutes, until the crust is hard and golden. Let stand for about 5 minutes before slicing and serving.

Yield: 4 servings

Deep-Dish Kale and Cheese Pizza

Kale is one of my favorite greens—a hearty, leafy vegetable loaded with vitamin A, calcium, and iron. It's a tasty green, one that you can really sink your teeth into. And best of all, it's a green really worth its space in a northern garden since it will hold up quite nicely under a snow cover. In fact, this green doesn't even reach full flavor until it has endured a frost or two.

Look for dark bluish green leaves; don't buy yellowed kale. The leaves and stalk should be firm, not limp. Kale that has sat around in the produce department will have developed a bitter taste. You can store kale for 3 to 4 days in a plastic bag in your refrigerator, but use it before it starts to yellow. Young leaves can be eaten stalk and all, but older leaves should be stripped from the tough center rib.

1 tablespoon (1 package) active dry yeast
2/3 cup warm water (105°F to 115°F)
1 tablespoon olive oil
1 teaspoon salt
1/2 cup cornmeal
1 to 1 1/2 cups all-purpose flour
6 cups shredded kale
3 cups ricotta cheese
2 to 4 garlic cloves, crushed
1 teaspoon fennel seeds
4 eggs, well beaten
1/4 cup all-purpose flour
1/2 cup slivered sun-dried tomatoes marinated in oil
5 to 6 slices provolone cheese

Sprinkle the yeast over the water in a large bowl. Stir to dissolve and let stand for about 5 minutes. Then mix in the remaining dough ingredients, reserving about 1/2 cup of the flour. Turn the dough out onto a lightly floured board and knead in as much of the remaining flour as needed to make a soft, elastic dough. Place the dough in a lightly oiled bowl, turning to coat the entire surface of the dough with the oil. Cover and let rise in a warm place for 45 to 60 minutes.

While the dough rises, prepare the filling. Steam the kale until limp, 3 to 5 minutes. Do not overcook. Drain well.

Combine the ricotta, garlic, fennel, eggs, and flour and mix with a spoon. Fold in the kale, then the sun-dried tomatoes. Set aside.

When the dough has doubled in bulk, punch it down. Stretch to fit over the bottom and up the sides of an oiled 10-inch springform pan or deep-dish pie pan.

Assemble the pizza while the oven preheats to 450°F.

Line the dough with the provolone cheese. Spoon on the ricotta-kale mixture.

Bake in the preheated oven for about 20 minutes. Then reduce the heat to 350°F and continue to bake for about 50 minutes, until the filling is set and a knife inserted near the center comes out clean. Allow to stand for about 10 minutes before removing the sides of the springform. Slice the pizza into wedges and serve.

Yield: 4 servings

PITA PIZZAS PROVENÇALE

This is the ultimate in lazy cooking—a pizza without the fuss of a yeast dough.

1 large garlic clove, minced
1/8 teaspoon hot pepper sauce
1/4 cup olive oil from marinated tomatoes
4 pita pockets (5 to 7 inches in diameter)
2 cups (8 ounces) shredded fontina cheese
1 cup chopped sun-dried tomatoes marinated in oil
1/2 cup sliced ripe olives
(2-ounce) can anchovy fillets, drained (optional)
2 to 3 tablespoons chopped fresh herbs (basil, savory, rosemary, sage) or 2 to 3 teaspoons dried

Place an ungreased baking sheet in the oven and preheat to 450°F.

Mix the garlic and hot pepper sauce with the oil. Brush onto both sides of each pita pocket, reserving about 1 tablespoon of the oil. Cover the pitas with half the cheese. Arrange the tomatoes, olives, anchovies, and herbs over the cheese. Top each with the remaining cheese; drizzle with the reserved oil. Place the pitas on the preheated baking sheet. Bake for 8 to 10 minutes, just until the pitas are crisp. Serve immediately.

Yield: 4 servings

Eggplant Feta Calzone

Basic Pizza Dough (page 41)
3 tablespoons olive oil
1 1/2 pounds eggplant, peeled and diced
1/4 teaspoon garlic powder
1/4 teaspoon salt
1/4 cup chopped sun-dried tomatoes marinated in oil
1 tablespoon chopped fresh basil or 1 teaspoon dried
1/4 pound feta cheese, crumbled
1/2 pound mozzarella, grated

Prepare the dough according to the recipe directions. While the dough rises, prepare the filling. Heat the oil in a large, heavy skillet. Add the eggplant, garlic powder, and salt and sauté for about 6 minutes, until the eggplant is completely tender. Remove from the heat and stir in the tomatoes, basil, and feta cheese. Set aside.

Preheat the oven to 500°F. Lightly oil a large baking sheet. When the dough has risen, punch it down. Stretch the dough, without tearing it, to an oval shape about 12 inches long. If you are a novice at stretching pizza dough, you may prefer to roll it on a lightly floured surface. Carefully transfer the dough to the baking sheet and stretch it to its final shape, about 15 inches long. Sprinkle half the grated cheese on half the dough to within 1/2 inch of the edge. Spoon the eggplant mixture on top of the cheese. Top with the remaining grated cheese. Brush the edge of the dough with water and fold the dough over the filling to form a half moon shape. Be sure the calzone is completely sealed. Bake for about 15 minutes, until the top is firm and golden. Serve warm.

Yield: 2 to 3 servings

Thyme-Flavored Focaccia

1 tablespoon (1 package) active dry yeast
1 1/3 cups warm water
2 tablespoons olive oil
Approximately 3 cups all-purpose flour
1 teaspoon salt
1/2 cup olive oil
3 tablespoons fresh thyme or 3 teaspoons dried
6 tablespoons chopped sun-dried tomatoes marinated in oil
Coarse (kosher) salt
Freshly ground black pepper

Sprinkle the yeast over the water in a large bowl. Stir to dissolve the yeast and let stand for about 5 minutes. Stir in the 2 tablespoons oil. Stir in the first 2 cups of flour along with the salt, then knead in the last cup or so. Knead on a well-floured board for about 8 minutes, until the dough is very soft and elastic but not sticky. Place the dough in a lightly oiled bowl, turning to coat the entire surface of the dough with oil. Cover and let rise in a warm place for about 1 hour.

While the dough rises, combine the 1/2 cup oil with the thyme. Heat gently for a few minutes. Then remove from the heat and let steep. Preheat the oven to 475°F. Lightly dust 2 baking sheets with cornmeal.

On a very lightly floured board, roll the dough into a long log, 3 to 4 inches in diameter. Cut the log into 6 equal-size pieces. By stretching and patting, form each piece of dough into a flat circle about 6 inches in diameter. Place on the prepared baking sheets. Generously brush each focaccia with the thyme-flavored oil. Sprinkle with the tomatoes and coarse salt and pepper. Bake for about 15 minutes, until golden brown. Serve hot.

Yield: 6 servings

Basil and Tomato Focaccia

In this focaccia, fresh basil and sun-dried tomatoes are kneaded into the dough. I like to form the dough into individual servings, but you can make 1 large, round or rectangular, pizza-size focaccia. While it is still warm from the oven, slice it into serving-size pieces as you would a pizza.

1 tablespoon (1 package) active dry yeast
1 1/3 cups warm water
2 tablespoons olive oil
1/2 cup minced fresh basil or 1/4 cup dried
1/2 cup minced sun-dried tomatoes marinated in oil
Approximately 3 cups all-purpose flour
1 teaspoon salt
1/2 cup olive oil
Coarse (kosher) salt
Freshly ground black pepper

Sprinkle the yeast over the water in a large bowl. Stir to dissolve the yeast and let stand for about 5 minutes. Stir in the 2 tablespoons oil, basil, and tomatoes. Stir in the first 2 cups of flour along with the 1 teaspoon salt, then knead in the last cup or so. Knead on a well-floured board for about 8 minutes, until the dough is very soft and elastic but not sticky. Place the dough in a lightly oiled bowl, turning to coat the entire surface of the dough with oil. Cover and let rise in a warm place for about 1 hour. Preheat the oven to 475°F. Lightly dust 2 baking sheets with cornmeal.

On a very lightly floured board, roll the dough into a long log, 3 to 4 inches in diameter. Cut the log into 6 to 8 equal-size pieces. By stretching and patting, form each piece of dough into a circle 6 inches in diameter. Place on the prepared baking sheets. Generously brush each focaccia with the 1/2 cup oil. Sprinkle with coarse salt and pepper. Bake for about 15 minutes, until golden brown. Serve hot.

Yield: 6 to 8 servings

Tomato Herb Rolls

These savory dinner rolls are perfect with soup. If you time everything just right, the wonderful aroma of these rolls will tease you as you wait for dinner. They are delicious fresh out of the oven.

1 cup milk
1 tablespoon oil
1 small onion, finely diced
1 tablespoon (1 package) active dry yeast
1 egg, slightly beaten
1/4 teaspoon powdered sage
1/2 teaspoon dried thyme
1 teaspoon salt
1/4 cup minced sun-dried tomatoes marinated in oil
3 to 4 cups all-purpose flour
1 egg yolk
1 tablespoon water

Scald the milk. Pour into a large mixing bowl and set aside to cool.

In a small frying pan, heat the oil. Add the onion and slowly sauté over medium heat until golden, about 4 minutes. Set aside.

When the milk has cooled to about 115°F, sprinkle in the yeast. Stir to mix; set aside until the mixture begins to foam, about 5 minutes. Then add the onion, egg, sage, thyme, salt, and tomatoes. Stir in the first 2 cups of flour, a cup at a time. Knead in the last 1 to 2 cups on a well-floured board. Knead until the dough is smooth and elastic, about 10 minutes.

Place the dough in a well-oiled bowl and turn to coat the entire surface of the dough with oil. Cover and set aside to rise in a warm place for 50 minutes.

Punch down the dough. Divide the dough into 12 pieces. Form the pieces into any roll shape you like. I prefer to make bow ties by rolling the dough into thin ropes, about 10 inches long, and tying each in a loose knot. You could also make round rolls by shaping the dough into balls. Or form cloverleafs by dividing each piece into thirds and rolling into little balls; place 3 little balls each in greased muffin cups to bake. For shapes other than a cloverleaf, set the rolls 1½ inches apart on a greased baking sheet. Cover and set aside to rise in a warm place for 45 minutes.

Lightly beat the egg yolk with the water and brush onto the rolls.

Preheat the oven to 375°F. Bake for about 20 minutes, until the rolls sound a little hollow and feel hard when tapped. Cool on a wire rack for a few minutes. Serve warm.

Yield: 12 dinner rolls

SUN-DRIED TOMATO AND PROVOLONE BREAD

This is a recipe for a savory fast bread—no rising time required.

2 tablespoons vegetable shortening, at room temperature
2 tablespoons sun-dried tomato oil
2 tablespoons sugar
2 garlic cloves, minced
2 large eggs, lightly beaten
1 1/4 cups buttermilk
2 1/2 cups all-purpose flour
2 teaspoons baking powder
1 1/4 teaspoons salt
1/2 teaspoon baking soda
1 cup grated provolone cheese
1/2 cup thinly sliced scallions
2 tablespoons minced fresh parsley
3/4 teaspoon coarsely ground black pepper
1/3 cup chopped sun-dried tomatoes
1/3 cup pine nuts, lightly toasted

Preheat the oven to 350°F. Grease 3 small loaf pans (5 1/2 inches by 3 1/8 inches by 2 1/4 inches).

In a small bowl, whisk together the shortening, oil, and sugar until smooth. Add the garlic, eggs, and buttermilk; whisk until the mixture is well combined. Set aside.

In a large bowl, sir together the flour, baking powder, salt, and baking soda. Mix in the cheese, scallions, parsley, pepper, and tomatoes. Make a well in the center and add the buttermilk mixture and pine nuts. Stir until just combined.

Pour the batter into the prepared loaf pans and smooth the tops. Bake in the preheated oven for 45 to 50 minutes, or until a toothpick inserted in the center of each loaf comes out clean. Set the pans on wire racks and allow the loaves to cool in the pans for about 10 minutes. Loosen the edges with a knife and turn the loaves out onto the racks to cool completely.

Well-wrapped in foil and chilled, the loaves will keep for up to 4 days.

Yield: 3 small loaves

Cheese and Tomato Biscuits

The secret to making light, flaky biscuits, I've found, is to use frozen butter and grate it into the flour as you would grate cheese. Mix it into the flour lightly, so that lumps of butter remain, and you will have wonderfully flaky biscuits. These buttery morsels are delicious with soups and salads.

3 1/2 cups sifted unbleached all-purpose flour
1 teaspoon salt
4 teaspoons baking powder
10 tablespoons frozen butter
1 1/2 cups milk
1 cup slivered sun-dried tomatoes marinated in oil
2 tablespoons freshly grated Parmesan cheese

Preheat the oven to 400°F.

Sift together the flour, salt, and baking powder. Grate the butter into the flour. Lightly mix with your fingertips to distribute the butter pieces throughout the flour. Marble-size lumps should remain. Mix in the milk to form a sticky dough.

Turn the dough out onto a lightly floured board and knead just a few times to make a smooth ball. Divide in half. Roll out half to a rectangle of approximately 8 inches by 12 inches, or roll out to a 10-inch square. The exact proportions are not important. Sprinkle the slivered tomatoes over the dough; then sprinkle the cheese over. Roll out the second half of the dough to the same dimensions as the first and fit on top. Trim away any uneven edges. Cut the dough into 2-inch squares. Carefully place the squares on ungreased baking sheets. Bake for 12 to 15 minutes, until golden brown. Cool briefly on wire racks before serving.

These are best served warm, but they can be held for several hours.

Yield: Approximately 28 biscuits

Hearty Cornbread

Corn, cheese, and tomatoes make this cornbread very filling—almost a meal in itself.

1 1/2 cups cornmeal
1/2 teaspoon baking soda
1/4 teaspoon black pepper
2 tablespoons minced fresh basil leaves or 2 teaspoons dried
4 tablespoons butter, melted
3 large eggs, well beaten
1 (17-ounce) can cream-style corn
1 cup sour cream
2 garlic cloves, minced
1 onion, finely chopped
1 1/2 cups (6 ounces) grated mozzarella
1/2 cup slivered sun-dried tomatoes marinated in oil

Preheat the oven to 375°F. Oil a 10-inch cast-iron skillet and place in the oven to preheat.

In a large bowl, mix together the cornmeal, baking soda, pepper, and basil.

In another bowl, mix together the butter, eggs, corn, sour cream, garlic, onion, mozzarella, and tomatoes.

Make a well in the center of the cornmeal mixture. Pour in the egg mixture. Stir just enough to combine. Pour the batter into the preheated skillet. Bake for 50 to 60 minutes, until the bread is golden and fairly firm, but not hard. Serve warm.

Yield: 6 to 8 servings

Red Basil Pesto

Sun-dried tomatoes are particularly potent in this extraordinary blend with opal basil, a red-hued variety. Serve this pungent, intense pesto as a sauce for pasta, fish, or fresh green beans. For an interesting recipe that combines the sauce with smoked turkey, *see page 65.*

1 1/2 to 2 cups fresh opal basil leaves
2 1/2 tablespoons minced sun-dried tomatoes marinated in oil
2 garlic cloves
2 tablespoons freshly grated Pecorino romano cheese
6 tablespoons freshly grated Parmesan cheese
1/3 cup pine nuts
1/2 cup olive oil
Salt and freshly ground black pepper

Mix the basil, tomatoes, garlic, cheese, and pine nuts in a food processor or blender. With the machine running, slowly add the olive oil. Season to taste with salt and pepper, and blend to the desired consistency. Let stand for 5 minutes before serving.

Yield: About 1 1/4 cups

Sun-Dried Tomato Sauce

The almonds give this rich, uncooked sauce an interesting texture, while the sun-dried tomatoes yield a mellow, sweet flavor. Like a stew, this superb sauce tastes even better the second day. It will keep in the refrigerator for 2 to 3 weeks, and it also freezes well.

1 garlic clove
1 small onion
2 cups sun-dried tomatoes marinated in oil
2 tablespoons olive oil
1/3 cup fresh lemon juice
1/4 cup freshly grated Parmesan cheese
1/3 cup slivered almonds, chopped
1/4 teaspoon dried thyme
1/4 teaspoon dried savory
1/2 teaspoon dried oregano
Salt and pepper to taste
1 1/2 cups chicken stock or broth

In a food processor fitted with a steel blade, chop the garlic and onion. Add the tomatoes and process carefully by pulsing the machine on and off until the tomatoes are just chopped. Do not overprocess or you will wind up with a paste.

Add the oil, lemon juice, Parmesan cheese, almonds, herbs, and salt and pepper. Process very briefly, just to mix, still preserving the rough texture. Pour the sauce into a bowl and stir in enough of the stock to make a loose, but chunky, tomato sauce. Set aside for at least 30 minutes to allow the flavors to blend. Taste and adjust the seasonings. Serve over hot pasta.

Yield: 4 servings

Mediterranean Pasta Sauce

Another sauce that requires no cooking, Mediterranean Pasta Sauce takes its rich flavor from olives, herbs, and tomatoes.

1 cup slivered sun-dried tomatoes marinated in oil
1/4 cup olive oil from marinated tomatoes
1 cup black olives, pitted and halved
1 cup loosely packed basil leaves or 1 tablespoon dried
1 tablespoon grated lemon peel
2 garlic cloves, minced
2 teaspoons freshly ground black pepper
1/4 pound mozzarella cheese, grated

Combine all the ingredients. If possible, make at least 1 hour before serving and set aside to allow the flavors to blend. Toss with fresh, hot pasta and serve.

Yield: 4 to 6 servings

Vermicelli with No-Cook Tomato Sauce

This is a rich pasta dish that is best served in small portions.

2 eggs
1/2 cup slivered sun-dried tomatoes marinated in oil
1/4 to 1/2 cup olive oil from marinated tomatoes
1/2 cup freshly grated Parmesan cheese
1/2 cup chopped fresh parsley
2 garlic cloves, minced
1 tablespoon lemon juice
8 ounces uncooked vermicelli or other thin spaghetti-shaped pasta
Salt and pepper to taste

In a serving bowl, whisk together the eggs, tomatoes, oil, cheese, parsley, garlic, and lemon juice. Set aside.

Cook the pasta in plenty of boiling salted water according to the package directions. Drain. While the pasta is still hot, add to the egg mixture. Toss well to coat the pasta with the sauce. Season with salt and pepper. Serve hot.

Yield: 4 first course servings

Pasta with Mushrooms and Greens

2 tablespoons olive oil
4 tablespoons butter
12 ounces fresh mushrooms, sliced (use a combination of domestic white, shiitake, porcini, oyster, or any other mushrooms)
2 large shallots, chopped
2 garlic cloves, chopped
2 tablespoons flour
2 cups chicken stock or broth
3/4 cup white wine
6 cups coarsely chopped greens (arugula and kale are both recommended)
3/4 cup chopped sun-dried tomatoes marinated in oil
Salt and pepper to taste
1 pound uncooked pasta (fusilli is recommended)

In a large skillet or saucepan, heat the oil and butter. Add the mushrooms, shallots, and garlic, and sauté until the mushrooms have yielded their liquid, about 5 minutes. Stir in the flour until well blended. Add the chicken stock and white wine, and simmer until slightly thickened, about 10 minutes. Add the greens and sun-dried tomatoes, and simmer for another 10 minutes to cook the greens and allow the flavors to blend. Season with salt and pepper.

Cook the pasta in plenty of boiling salted water until just tender. Drain and pour into a large serving bowl. Add the sauce and toss. Serve at once.

Yield: 4 servings

Pasta with Green Beans

12 ounces fresh green beans (2 large handfuls)
7 tablespoons olive oil from marinated tomatoes
4 garlic cloves, finely minced
2 sprigs fresh savory
1/2 cup chopped fresh parsley
6 sun-dried tomatoes marinated in oil, julienned
12 ounces uncooked pasta (smaller kinds, such as tagliarini, spaghetti, or vermicelli work best)
Parmesan cheese

Preheat a large quantity of water for the pasta. Trim the ends from the green beans. Blanch them for 1 minute in the water prepared for the pasta. Remove from the water and drain well.

Heat the olive oil in a skillet. Add the garlic and simmer gently for 2 minutes. Add the savory, parsley, green beans, and tomatoes. Heat through.

Cook the pasta until just tender. Drain; place on a large serving dish. Toss the green bean mixture with the pasta and serve. Pass Parmesan cheese to sprinkle on top.

Yield: 4 servings

Fettuccine alla Francesca

When sun-dried tomatoes started turning up on menus in Vermont, I wrote a story about them for the Burlington Free Press. As part of the article, I collected several recipes from local restaurateurs, including this sinfully rich one I've adopted from Scott Vineberg of Francesca's Restaurant in Shelburne, Vermont.

4 cups cream
10 tablespoons butter
2 teaspoons minced garlic
2 teaspoons freshly ground black pepper
Pinch nutmeg
Salt to taste
3/4 pound uncooked fettuccine
1/2 cup fresh or frozen peas
1/2 cup sliced mushrooms
1/4 cup slivered sun-dried tomatoes
Freshly grated Parmesan cheese

Combine the cream, butter, garlic, pepper, nutmeg, and salt in the top of a double boiler. Heat thoroughly.

Cook the pasta in plenty of boiling salted water until just tender, then drain.

Add the peas, mushrooms, tomatoes, and cooked pasta to the cream sauce and toss. Serve at once, with freshly grated cheese on the side.

Yield: 4 servings

Orzo with Sun-Dried Tomatoes

This is a great side dish with roast chicken. If you like, make this early in the day and heat for about 30 minutes before serving.

3 cups uncooked orzo
4 tablespoons butter
2 large garlic cloves, minced
1 (14-ounce) can artichoke hearts, drained and quartered
1/2 cup slivered sun-dried tomatoes marinated in oil
1/4 cup chopped fresh parsley
Freshly ground black pepper to taste
2 tablespoons freshly grated Parmesan cheese

Cook the orzo in plenty of boiling salted water for 8 to 10 minutes. Drain and pour into a 2-quart casserole. Preheat the oven to 350°F.

In a medium-size skillet, melt the butter. Add the garlic and simmer for about 1 minute. Add the artichokes and sauté for about 1 minute. Stir in the tomatoes and remove from the heat.

Stir the artichoke mixture into the orzo along with the parsley. Season with black pepper. Smooth the top and sprinkle with the cheese. Bake in the preheated oven for 15 to 20 minutes, until the top is golden. Serve hot.

Yield: 6 to 8 side dish servings

Vermicelli with Red Pesto

This pasta, with its rich flavors of pesto and smoked turkey, is best savored in small portions as a first course.

1/2 pound uncooked vermicelli
2 tablespoons heavy cream
About 1 tablespoon hot water in which the pasta was cooked
1/2 cup Red Basil Pesto (page 57)
1/2 cup minced smoked turkey
Freshly grated Parmesan cheese
4 tablespoons very finely slivered sun-dried tomatoes marinated in oil

Cook the pasta in plenty of boiling salted water according to the package directions.

Stir the cream and 1 tablespoon of the hot cooking water into the pesto. Drain the pasta and return it to the hot pot. Toss with the pesto and smoked turkey. Serve garnished with freshly grated Parmesan cheese and slivered tomatoes.

Yield: 4 first course servings

Pasta with Red Pesto and Sausage

1 pound hot Italian sausage, sliced
4 garlic cloves, minced
1 (14-ounce) can artichoke hearts, quartered
1/4 cup slivered sun-dried tomatoes marinated in oil
1/2 cup red wine
12 ounces uncooked fettuccine
1/2 cup Red Basil Pesto (page 57)
Freshly grated Parmesan cheese

Brown the sausage over medium-high heat for about 5 minutes. Add the garlic, then the artichokes, tomatoes, and wine. Simmer over medium heat for 8 to 10 minutes.

Cook the fettuccine in plenty of boiling salted water according to the package directions. Drain and place on a serving platter. Toss the pasta with the Red Basil Pesto and spoon the artichoke sauce on top. Serve at once, passing the Parmesan cheese at the table.

Yield: 4 servings

SONOMA PASTA

8 thick-cut slices smoked bacon
12 ounces uncooked fettuccine
2 tablespoons minced garlic
12 medium-size basil leaves, slivered, or 2 teaspoons dried
1 cup slivered sun-dried tomatoes marinated in oil
1 cup whole black olives, pitted
8 ounces goat cheese, crumbled
Salt and pepper to taste

Cut the bacon into 3/4-inch pieces; brown well in a heavy skillet. Remove the bacon from the skillet and drain on paper towels. Reserve the bacon drippings.

Begin cooking the pasta in plenty of boiling salted water according to the package directions.

Add the garlic to the skillet and sauté for 2 to 3 minutes. Add the basil leaves and tomatoes and sauté briefly until limp, about 1 minute. Add the olives and heat through. Add the cheese and bacon. Season with salt and pepper.

Drain the pasta, add immediately to the sauce, and toss well. (The cheese should be only partially melted.) Serve at once.

Yield: 4 servings

Spaghetti and Shrimp in Garlic Sauce

Here's a lovely sauce that can be prepared in less time than it takes to cook the spaghetti—leaving you time to toss together a green salad for a wonderful meal.

6 tablespoons butter
6 tablespoons olive oil from marinated tomatoes
2 large garlic cloves, minced
6 tablespoons chicken stock
1 1/3 pounds shrimp, peeled and deveined
2/3 cup sun-dried tomatoes marinated in oil
1/8 teaspoon crushed red pepper flakes
2/3 cup chopped fresh parsley
1 pound uncooked spaghetti
2 tablespoons butter

While the cooking water for the spaghetti heats, make the sauce. Combine the 6 tablespoons butter and olive oil over medium heat. Add the garlic and simmer for about 2 minutes. Add the chicken stock and heat. Add the shrimp and simmer until pink and firm, about 3 minutes. Stir in the tomatoes, red pepper, and parsley; keep warm.

Cook the spaghetti until just tender. Drain. Toss with the remaining 2 tablespoons butter. Pour onto a large warmed serving platter. Pour the shrimp mixture over the spaghetti and toss. Serve at once.

Spaghetti and Scallops in Garlic Sauce. Substitute 1 pound scallops for the shrimp and proceed with the recipe as above, simmering the scallops for 5 to 6 minutes.

Yield: 4 servings

Linguine with Scallops

1 1/2 pounds scallops
1 1/2 cups white wine
1/2 cup water
2 bay leaves
1 teaspoon thyme
12 ounces uncooked linguine
1/4 cup butter
2 shallots, minced
1/2 cup slivered sun-dried tomatoes marinated in oil
1/2 cup heavy cream
Freshly ground black pepper to taste

If you are using sea scallops, slice them into thirds. In a large saucepan, combine the wine, water, bay leaves, and thyme. Bring to a boil, then reduce the heat to a gentle simmer. Add the scallops and poach in the just-simmering liquid for 5 to 6 minutes, until firm and tender. Remove from the liquid and keep warm. Strain out 1 1/2 cups poaching liquid into a measuring cup.

Cook the linguine in plenty of boiling salted water until just tender. Drain; keep warm.

In a large skillet, melt the butter. Add the shallots and sauté for about 1 minute. Add the reserved poaching liquid, tomatoes, and cream. Bring to a boil and cook until the sauce is somewhat reduced and slightly thickened. Stir in the scallops. Season with pepper.

Pour the linguine onto a serving platter. Pour the scallops and sauce over and toss gently. Serve at once.

Linguine with Shrimp

Substitute 1 1/2 pounds medium shrimp (peeled and deveined) for the scallops, and proceed with the recipe as above, poaching the shrimp for 3 to 4 minutes.

Yield: 4 servings

White Lasagna

This is a very rich dish. Small servings are guaranteed to suffice.

3/4 pound uncooked lasagna noodles
1/2 cup butter
1/2 cup flour
4 cups milk
1/2 cup white wine
Salt and pepper to taste
Nutmeg
1 pound hot Italian sausage
1 pound frozen chopped spinach, defrosted
3/4 cup slivered sun-dried tomatoes marinated in oil
3/4 pound provolone cheese, sliced (about 18 slices)
2 tablespoons freshly grated Parmesan cheese

Cook the lasagna noodles in plenty of boiling salted water until just tender. Drain and set aside.

Melt the butter in a saucepan. Whisk in the flour and cook over medium heat until all lumps are dissolved, about 2 minutes. Add the milk and wine and cook until the sauce is thickened, about 10 minutes. Add salt and pepper and nutmeg to taste.

Remove the sausage meat from the casings and crumble. Brown the meat in a skillet over medium heat, about 8 minutes. Remove from the skillet and drain on paper towels.

Squeeze the defrosted spinach to be sure all excess moisture is removed.

Now you are ready to assemble the lasagna. Preheat the oven to 350°F.

Spoon a small amount of sauce into the bottom of a 9-inch by 13-inch baking dish. Cover with a layer of noodles. Cover the noodles with a layer of spinach (use all the spinach). Sprinkle with about 1/4 cup of the tomatoes. Cover with a layer of provolone cheese. For the next layer, use noodles, then sauce, then sausage, then 1/4 cup tomatoes, then provolone cheese. Follow with noodles, sauce, tomatoes, and a final layer of provolone cheese. Cover with a layer of noodles, then a layer of sauce, and sprinkle with Parmesan cheese.

Bake for about 30 minutes, until completely heated through. Allow to stand for 5 to 10 minutes before serving.

Yield: 6 to 8 servings

Mussels in Sun-Dried Tomato Broth

A wonderfully simple dish, richly flavored with sun-dried tomatoes. Serve with a loaf of crusty French bread for dipping in the broth.

4 pounds mussels in shells
1 large shallot, minced
3 bay leaves
1 teaspoon black peppercorns
1 tablespoon chopped fresh oregano or 1 teaspoon dried
1 cup white wine
1 cup water
2 to 3 tablespoons sun-dried tomato puree (see page 8)
1/2 cup thinly sliced scallions
1/2 yellow pepper, finely diced
4 tablespoons chopped fresh parsley

Scrub the mussels and remove the beards. Place in a large stockpot. Add the shallot, bay leaves, peppercorns, oregano, wine, and water. Cover tightly, bring to a boil, and steam for about 5 minutes, or until most of the shells are opened. Discard any shells that remain closed.

Pour out the broth; cover the pot to keep the mussels warm. To remove any sand, strain the broth through a double thickness of cheesecloth that has been moistened under tap water. Bring the broth to a gentle boil; then stir in the tomato puree until it dissolves. Add the scallions and yellow pepper and remove from heat.

Divide the mussels among 4 soup bowls. Sprinkle each bowl with a tablespoon of parsley. Pour the broth over the mussels and serve at once.

Yield: 4 servings

Shrimp and Vegetables in White Wine

2 tablespoons butter
2 tablespoons olive oil from marinated tomatoes
1 red or yellow pepper, julienned
2 carrots, julienned
4 scallions, including green tops, diced
1 (14-ounce) can artichoke hearts packed in water, drained and sliced
1 1/4 cups fresh or frozen peas
1 1/2 pounds medium shrimp, peeled and deveined
1/2 cup slivered sun-dried tomatoes
1/2 cup white wine
Salt and pepper to taste
Hot cooked rice

In a large, heavy skillet, heat 1 tablespoon of the butter and 1 tablespoon of the oil over medium high heat. Add the pepper, carrots, and scallions; sauté until just limp, about 2 minutes. Stir in the artichoke hearts and peas. Remove the vegetables to a bowl and keep warm.

Heat the remaining 1 tablespoon oil and 1 tablespoon butter in the skillet. Add the shrimp and sauté for 1 minute. Add the tomatoes and wine and continue to cook until the shrimp are pink and firm, 4 to 5 minutes. Toss the vegetables with the shrimp in the skillet. Season to taste with salt and pepper. Serve at once over hot cooked rice.

Yield: 4 to 5 servings

Trout Stuffed with Lemon Rice

I love the flavor of fresh trout. Unfortunately, even the most carefully prepared fish cannot be served without a warning to watch for bones.

3 tablespoons butter
3 tablespoons minced onion
1 garlic clove, minced
1 cup uncooked white rice
2 cups chicken stock or bouillon
Salt to taste
1/2 cup chopped fresh parsley
Juice and grated rind of 1 lemon
3 tablespoons slivered sun-dried tomatoes marinated in oil
1 boned whole trout (about 1 1/2 pounds)
1 tablespoon butter

In a large heavy skillet, melt 3 tablespoons of the butter. Add the onion, garlic, and rice; sauté over medium-low heat until the rice is golden, about 5 minutes. Add the chicken stock and salt. Stir, cover, and cook until all the liquid is absorbed, about 15 minutes. Stir in the parsley, lemon rind, and tomatoes. Remove from the heat.

Preheat the oven to 400°F.

Lightly butter a baking-serving dish. Place the fish in the dish and butter both sides, using the remaining 1 tablespoon butter. Stuff some rice into the cavity of the trout. Spoon the remaining rice around the fish. Cover the rice with aluminum foil (to prevent it from drying out in the oven), leaving the fish exposed. Bake for about 10 minutes per inch thickness of the stuffed fish. (Measure with a ruler at the thickest part of the fish; it should be about 1 1/2 to 2 inches thick.) Bake until the flesh feels firm and flakes easily when tested with a fork. Serve at once.

Yield: 2 servings

Baked Stuffed Monkfish Fillets

Monkfish has been called "the poor man's lobster" Its sweet-tasting flesh combines nicely with the filling in this recipe. If monkfish isn't available, substitute another flavorful white fleshed fish, such as flounder, ocean perch, or sea bass.

1 celery stalk, minced
1 carrot, minced
1 small onion, minced
2 tablespoons chopped capers
1/2 cup chopped sun-dried tomatoes marinated in oil
1 cup dry bread crumbs
Approximately 1/3 cup white wine
2 to 2 1/2 pounds monkfish fillets
1/4 to 1/3 cup bread crumbs
Approximately 3 tablespoons olive oil from marinated tomatoes

Preheat the oven to 350°F.

In a mixing bowl, combine the celery, carrot, onion, capers, tomatoes, and 1 cup bread crumbs. Moisten with the wine. Lightly oil a shallow baking dish. Place half the fillets in the dish. Distribute the bread crumb mixture over the fillets, patting it into place. Top with the remaining fillets. Sprinkle the remaining bread crumbs on top and drizzle with the oil.

Bake for 20 to 30 minutes, until the fish feels firm and flakes easily with a fork. Allow to rest for 5 minutes before serving.

Yield: 4 servings

Fish Fillets in Wine Sauce

Just about any fish is good with this creamy sauce. Sea bass is my favorite, but haddock, turbot, and red snapper are all good choices.

4 fish fillets, about 6 ounces each
4 teaspoons butter
Salt and pepper to taste
1 tablespoon butter
1 large shallot or 1 small onion, diced
1/2 cup white wine
1/2 cup dry sherry
3 tablespoons minced sun-dried tomatoes marinated in oil
1 cup heavy cream

Preheat the oven to 400°F.

Place each fillet on an 11-inch piece of aluminum foil. Dot each with 1 teaspoon butter and season lightly with salt and pepper. Fold the foil over and seal the edges to form a packet. Place the packets on a baking sheet. Bake on the middle oven rack for 15 minutes.

Meanwhile, melt 1 tablespoon butter in a skillet. Add the shallot and sauté for 2 minutes. Add the wine and sherry; bring to a boil. Boil until the liquid is reduced to about 1/4 cup, about 5 minutes. Stir in the tomatoes. Add the cream and boil over high heat until the sauce is reduced to about 3/4 cup, about 5 minutes.

Remove the fish from the foil and place on warmed serving plates. Pour the sauce over the fish and serve at once.

Yield: 4 servings

Quick Baked Chicken with Tomatoes

8 chicken thighs or chicken breast halves
1 cup minced sun-dried tomatoes marinated in oil
Juice of 1 lemon
Salt and pepper to taste
1 tablespoon dried rosemary

Preheat the oven to 350° F.

Place the tomatoes in the bottom of an ungreased baking dish large enough to hold all the chicken pieces in a single layer. Lay the chicken on top of the tomatoes. Squeeze the juice of the lemon all over the chicken. Sprinkle with salt, pepper, and rosemary. Cover and bake until the chicken is tender, 40 to 45 minutes. Before serving, spoon the pan juices over the top of the chicken.

Yield: 4 servings

Quick Chicken Paella

Here's an easily made casserole that is reminiscent of the rich paellas of Spain and southern France. Serve with a green salad and French bread.

1/4 cup olive oil
2 garlic cloves, minced
1/2 cup chopped onion
1 frying chicken, quartered
1 cup slivered sun-dried tomatoes marinated in oil
4 cups chicken stock or broth
1 teaspoon ground saffron
1 dried bay leaf, crushed
1 cup uncooked rice
2 green peppers, julienned
1/4 cup fresh or frozen green peas

Heat the oil in a large skillet. Add the garlic and onion and sauté until transparent, 2 to 3 minutes. Remove and reserve for later. Add the chicken and fry until lightly browned, about 10 minutes. Return the garlic and onion to the pan along with the tomatoes. Add the chicken stock. Bring to a boil and simmer for 5 minutes. Add the saffron, bay leaf, rice, and peppers. Pour into a 3 1/2-quart or 4-quart casserole and cover.

Bake for 20 minutes. Stir in the peas and cover for 1 minute to allow the peas to heat through. Serve hot.

Yield: 4 servings

Chicken-Vegetable Packets

To make these incredibly flavorful chicken and vegetable packets you will need kitchen parchment, which you can find in many specialty food stores. If you can't find kitchen parchment, substitute aluminum foil.

2 cups assorted julienned vegetables (zucchini, carrots, red bell peppers, crookneck yellow squash)
4 boned and skinned chicken breast halves (about 6 ounces each)
1/2 cup chopped sun-dried tomatoes marinated in oil
1/4 cup lemon juice
2 teaspoons dried basil
4 teaspoons minced garlic
Salt and pepper to taste

Preheat the oven to 400°F. Cut four 12-inch circles from the kitchen parchment. Fold each circle in half; then unfold. Place 1/2 cup of the vegetables in the center of each circle, next to the fold. Top with a chicken breast half; sprinkle with a portion of the remaining ingredients. Refold the paper over the chicken so that the cut edges meet. Fold and roll the edges up all the way around to seal.

Place the packets on a baking sheet; bake for 20 to 25 minutes until the packets are browned and puffed. Transfer to plates; cut open to serve.

Yield: 4 servings

Chicken and Mushroom Sauté

Here's a dish to make when you want good food fast. Serve with rice and a green salad.

2 tablespoons olive oil
1/4 cup minced shallots
8 ounces mushrooms, sliced
1 red pepper, diced
1 small zucchini, quartered and sliced
1 teaspoon dried thyme or more to taste
2 tablespoons olive oil
4 boneless chicken breasts, cut in sixths or eighths
1 teaspoon dried thyme or more to taste
1 tablespoon balsamic vinegar
3/4 cup chopped sun-dried tomatoes marinated in oil
Salt and pepper to taste

Heat 2 tablespoons of the oil in a large skillet. Add the shallots and mushrooms and sauté for about 2 minutes. Add the pepper, zucchini, and 1 teaspoon thyme and sauté until the vegetables are just barely tender, about 2 more minutes. Remove from the skillet and keep warm.

Add the remaining 2 tablespoons oil to the skillet. Then add the chicken and the remaining 1 teaspoon thyme; sauté until the chicken is done, 12 to 15 minutes. Remove the chicken from the skillet and combine with the vegetables.

To the juices in the skillet, add the vinegar. Cook for a minute, stirring up any browned particles clinging to the bottom of the pan. Return the chicken and vegetables to the pan, along with the tomatoes. Toss together for a minute or two to blend the flavors. Season with salt and pepper. Serve at once.

Yield: 4 servings

Chicken in White Wine Sauce

3 tablespoons butter
4 skinless, boneless chicken breast halves (about 1 1/4 pounds), cut into sixths or eighths
1/2 teaspoon salt
1/4 teaspoon pepper
1 large shallot, minced
2/3 cup heavy cream
1/2 cup white wine
1/8 teaspoon marjoram
1/2 cup slivered sun-dried tomatoes marinated in oil

Cut the chicken breasts on the diagonal into six equal-size pieces. Melt the butter in a heavy skillet. Add the chicken slices and sprinkle with salt and pepper. Sauté over moderate heat until the chicken is opaque, 4 to 5 minutes. Using a slotted spoon, remove the chicken to a serving platter and keep warm.

Add the shallot to the skillet and sauté for 1 minute. Add the cream, wine, and marjoram. Bring to a simmer over moderate heat and cook, uncovered, stirring occasionally, until the sauce thickens slightly, about 5 minutes. Return the chicken to the skillet, add the tomatoes, and simmer until heated through, 2 to 3 minutes. Serve hot.

Yield: 4 servings

Chicken Scallopine alla L'Esprit

4 chicken breast halves, skinned and boned
4 slices fontina or Swiss cheese
4 ounces prosciutto, chopped
1/2 cup chopped sun-dried tomatoes
4 teaspoons Parmesan cheese
2 tablespoons olive oil
2 tablespoons minced shallots
1 cup chicken stock
1/4 cup white wine

Pound the chicken breasts between sheets of waxed paper until approximately 1/4 inch thick. Set aside 2 tablespoons of the prosciutto and 1/4 cup of the tomatoes for the sauce. Divide the remaining prosciutto, tomatoes, fontina, and Parmesan into 4 portions and place each on the center of a chicken breast. Roll into a cylinder and tie with string.

Heat the olive oil over medium-high heat, add the chicken, and brown. Remove the chicken and keep warm. Add the shallots and sauté for 2 minutes. Add the wine and stock and scrape up any browned bits from the bottom of the pan. Return the chicken to the pan, cover, and cook for 20 minutes over low heat. Remove the chicken and keep warm. Cook the sauce over high heat until it thickens and coats the back of a spoon. Add the reserved prosciutto and tomatoes and heat for about 5 minutes. Pour the sauce over the chicken breasts and serve.

Yield: 4 servings

Portuguese Kale Soup

Sun-dried tomatoes are not traditional in Caldo Verde, or Portuguese Kale Soup. But I find they marry well with the ingredients and add a nice flavor to the soup. This soup makes a wonderful wintertime meal.

1/2 pound linguica or chorizo sausage (or any garlicky smoked sausage), sliced
4 medium-size potatoes, peeled and diced
6 cups chicken stock
1 onion, diced
4 garlic cloves, crushed
4 to 5 cups chopped kale
1/2 cup chopped sun-dried tomatoes marinated in oil
Salt and pepper to taste

Combine the sausage with water to cover in a saucepan. Bring to a boil and boil gently for 5 minutes. Drain off the water and set the sausage aside.

Combine the potatoes with water to cover in a saucepan. Boil until tender, about 8 minutes. Drain, and briefly mash the potatoes with a potato masher. The texture should be uneven and lumpy.

In a large saucepan, combine the chicken stock, onion, and garlic, and bring to a boil. Reduce the heat and stir in the potatoes, then the sausage and kale. Simmer for about 15 minutes. Add the tomatoes and salt and pepper, if needed, and serve.

Yield: 4 servings

Chicken Soup with Greens

3-pound chicken, cut up (or 3 pounds chicken parts; dark meat yields a richer broth)
3 quarts water
2 onions, quartered
3 celery stalks, quartered
2 carrots, quartered
1 bunch fresh parsley
Salt to taste
1 cup uncooked alphabets or tubettini or other small pasta
3/4 cup chopped sun-dried tomatoes marinated in oil
3 to 4 cups arugula (or kale)

Combine the chicken, water, onions, celery, carrots, and parsley in a large soup pot. Bring to a boil, cover, and reduce the heat. Simmer for about 1 hour, until the chicken is tender.

Remove the chicken and let cool. Meanwhile, strain the soup and discard the vegetables.

Remove the chicken from the bones and cut into bite-size pieces. Reserve about 2 cups of meat for the soup and set aside the rest to use in another dish.

Return the strained soup to the pot and season to taste with salt. Bring to a boil. Add the pasta and boil vigorously until done, about 10 minutes. Reduce the heat and add the tomatoes and greens. Simmer until the greens are cooked, about 10 minutes. Serve at once.

Yield: 4 servings

Mary's Smoked Chicken Quiche

I came across this rich brunch recipe at a restaurant called Mary's in Bristol, Vermont. The tomatoes, smoked chicken, and avocado make a wonderful combination.

2 1/4 cups heavy cream
8 ounces Swiss cheese, grated
Unbaked 9-inch pie crust
8 ounces smoked chicken breast, diced
1 avocado, peeled and diced
1/4 cup chopped sun-dried tomatoes marinated in oil
6 eggs
1/2 teaspoon dried thyme
1/4 teaspoon cayenne pepper
Dash nutmeg

Preheat the oven to 375°F. Warm the cream while you prepare the other ingredients.

Place half the cheese in the prepared pie shell. Add the chicken, avocado, and tomatoes. Cover with the remaining cheese.

Combine the eggs, thyme, cayenne, and nutmeg in a bowl; beat well. Slowly add the warmed cream. Pour the mixture into the pie shell. Bake the quiche for 45 minutes. Remove from the oven and let sit for about 20 minutes to allow the custard to set. Serve warm.

Yield: 6 to 7 servings

Tomato and Fontina Quiche

Although other cheeses, especially Swiss, work well with this recipe, it is really worth the effort to seek out a high-quality fontina cheese. The nutty, sweet flavor of fontina and its terrific texture when melted are a wonderful complement to the salty, chewy tomatoes.

Unbaked 9-inch pie shell
1/2 cup chopped sun-dried tomatoes marinated in oil
1 cup grated Italian fontina cheese
1/3 cup thinly sliced scallions
1 1/4 cups half-and-half
2 large eggs
White pepper to taste

Prick the bottom of the pie shell with a fork. Chill for 30 minutes or freeze for 15 minutes.

Preheat the oven to 425°F. Line the pie shell with waxed paper and fill with pie weights or beans. Bake for 10 minutes, remove the paper and weights, and continue to bake for another 5 to 6 minutes, or until golden. Reduce the oven temperature to 375°F and cool the pie shell while you prepare the filling.

In a bowl, toss together the tomatoes, cheese, and scallions. Sprinkle the mixture evenly in the pie shell. In a bowl, whisk together the half-and-half, eggs, and pepper. Pour into the shell and bake for 30 to 35 minutes, or until a knife inserted 1/2 inch from the edge comes out clean. The custard in the center of the quiche won't be fully set, but it will continue to cook after being removed from the oven. Let stand for about 20 minutes. Serve warm.

Yield: 4 servings

Spinach Ricotta Pie

4 eggs, well beaten
1/2 cup unbleached all-purpose flour
2 (1-pound) packages frozen chopped spinach, thawed and drained
1 pound ricotta cheese
1/4 cup chopped fresh basil or 1 tablespoon dried
1/4 cup chopped shallot
3/4 cup chopped sun-dried tomatoes marinated in oil
3 tablespoons pine nuts
8 sheets filo dough
1/4 cup butter
2 tablespoons olive oil from marinated tomatoes

Preheat the oven to 350°F.

Beat together the ricotta, eggs, and flour. Stir in the basil and shallots. Squeeze the spinach to remove any excess liquid. Fold into the ricotta mixture. Then fold in the tomatoes and pine nuts.

Combine the butter and oil and heat just until the butter is melted. Generously brush the mixture inside a 10-inch springform pan. Fit a sheet of filo into the pan, draping the excess dough over the sides. Brush with the oil and butter mixture. Repeat until you have lined the pan with 5 sheets of buttered filo. Spoon the filling into the pan. Fold the extra filo over the filling. Cut 3 sheets of filo into a circle slightly larger than the pan. Place a circle on top of the pie, rocking the dough down inside the pan. Brush with the oil and butter mixture. Repeat with the remaining sheets.

Bake for 30 to 35 minutes, until the top is golden and a knife inserted near the center comes out clean. Let sit for at least 10 minutes before removing the sides of the pan and serving.

Yield: 6 servings

Tomato and Avocado Omelet

In each bite you get the contrasting flavor and texture of the creamy avocado and the chewy, rich tomatoes. A wonderful combination for a Sunday brunch or a quick supper.

2 teaspoons olive oil from marinated tomatoes
1 tablespoon minced shallot
2 eggs, well beaten
1/4 cup finely diced ripe avocado
2 sun-dried tomatoes marinated in oil, slivered

Heat the oil in an omelet pan over medium high heat. Add the shallots and sauté until limp, about 1 minute. Pour the eggs into the middle of the pan. As the egg sets, use a wooden spatula to push the cooked egg toward the center of the pan, allowing the uncooked egg to run to the outside. When the egg is mostly set, sprinkle with the avocado and tomatoes, reserving a few slivers for a garnish. Shake the omelet forward so the front rides up the side of the pan. Fold the front third of the omelet back over the filling. Slide the omelet onto a warmed plate and fold the remaining third of the omelet over the filling. Garnish with the reserved slivers of tomato and serve.

Yield: 1 serving

Tomato-Eggplant Stew

This is a very rich side dish. The richness is derived from the generous 2 tablespoons of sun-dried tomato puree as well as 3/4 cup of sun-dried tomatoes. The stew pairs well with simple meat dishes, such as broiled chicken. But you can also serve it as a vegetarian main dish with couscous or rice to accompany it.

1/4 cup olive oil
1 large onion, diced
1 pound eggplant, peeled and diced
1 teaspoon dried rosemary
1 green pepper, diced
2 celery stalks, sliced
1 1/2 cups sliced mushrooms
2 tablespoons sun-dried tomato puree (see page 8)
3/4 cup sun-dried tomatoes marinated in oil
Salt and pepper to taste
Hot pepper sauce to taste (optional)

Heat the oil in a large skillet. Add the onion and sauté until limp, about 2 minutes. Add the eggplant and rosemary and stir to coat the eggplant in the oil. Cover and cook over medium heat for about 5 minutes, stirring occasionally, until the eggplant is quite soft. Add the green pepper, celery, and mushrooms; sauté for about 5 minutes. The eggplant will be very soft, but the green vegetables will still be slightly crisp. Stir in the tomato puree, then the tomatoes. Add salt and pepper and hot pepper sauce. Cook for an additional 5 minutes over low heat to allow the flavors to blend. Serve hot.

Yield: 4 servings

Risotto with Sun-Dried Tomatoes

Arborio is the rice variety used by Italians to achieve the creamy, yet firm, texture of risotto. Surprisingly, Uncle Ben's Converted Rice will yield similar results, with each grain separate, yet bound together with a creamy consistency. It is very important to cook the rice slowly and gently; the total cooking time should be about 25 minutes.

This simple dish makes a fine accompaniment for a roast.

4 1/2 cups chicken broth or stock
4 tablespoons butter
2 garlic cloves, minced
1/4 cup minced shallot
1 1/2 cups Arborio rice or Uncle Ben's Converted Rice
1/4 cup dry white wine
1/4 cup chopped sun-dried tomatoes marinated in oil
1/4 cup freshly grated Parmesan cheese
1/4 cup chopped fresh parsley (optional)
Salt and pepper to taste

Heat the broth and maintain at a simmer. In another saucepan, melt 2 tablespoons of the butter. Add the garlic and shallots and sauté until limp, about 2 minutes. Add the rice and stir to coat with the butter. Add the wine and cook, stirring occasionally, until the wine is absorbed. Add 1/4 cup of the broth and cook over medium heat until the broth is absorbed. Continue to cook and add 1/4 cup of the broth at a time, stirring occasionally, until all the broth has been absorbed.

When the broth is all absorbed, stir in the remaining 2 tablespoons butter. Then stir in the tomatoes, cheese, and parsley. Season with salt and pepper. Serve at once.

Yield: 4 servings

Gratin of Zucchini and Carrot

When you want to make a vegetable dish that is a little special . . .

2 small zucchini, cubed
4 carrots, cubed
4 tablespoons butter
2 garlic cloves, minced
1/2 cup slivered sun-dried tomatoes marinated in oil
4 tablespoons freshly grated Parmesan cheese

Preheat the oven to 350°F.

Blanch the zucchini and carrots in boiling water for 1 minute. Do not overcook. Drain well.

Melt the butter and add the garlic. Simmer for about 1 minute.

In a gratin dish or low-sided casserole, toss together the vegetables and sun-dried tomatoes. Pour over about 3 tablespoons of the butter and toss gently. Sprinkle the Parmesan over the top of the vegetables and drizzle the remaining butter on top. Bake for 20 minutes. Serve at once.

Yield: 4 to 6 servings

Baked Vegetable Medley

2 medium-size zucchini, thinly sliced
1 medium-size onion, thinly sliced
1 teaspoon Italian seasoning
Salt and pepper to taste
1/2 cup sun-dried tomatoes marinated in oil, slivered
8 ounces mozzarella cheese, grated
2 medium-size yellow squash, thinly sliced

Preheat the oven to 350°F.

Lightly oil an 8 1/2-inch by 11-inch baking dish with oil from the tomatoes. Cover the bottom of the dish with all the zucchini. Add half the sliced onion and sprinkle with 1/2 teaspoon Italian seasoning and salt and pepper. Next layer on half the tomatoes, then half the cheese, then the yellow squash, then the remaining onions, the remaining 1/2 teaspoon Italian seasoning, and more salt and pepper. Finish the layering with the remaining dried tomatoes and grated cheese. Cover the dish with foil. The dish can be prepared a day ahead at this point and refrigerated for 24 hours. Bring to room temperature before baking.

Bake for 20 minutes. Uncover and bake for an additional 10 to 15 minutes, just until the cheese starts to brown. Serve hot.

Yield: 8 servings

Baked Spaghetti Squash with Tomatoes

2-pound spaghetti squash
2 tablespoons olive oil
1 green pepper, diced
1 cup diced celery
2 shallots, minced
1 cup slivered sun-dried tomatoes marinated in oil
1 teaspoon powdered rosemary
5 tablespoons freshly grated Parmesan cheese

Fill a large pot with water and heat to boiling. Add the squash and boil for 20 to 30 minutes, until easily pierced with a fork. Drain and cool. Preheat the oven to 350°F.

While the spaghetti squash cools, heat the oil in a large skillet. Add the green pepper, celery, and shallot and sauté until limp, about 2 minutes.

Cut the squash in half, and discard the seeds and fibers to which the seeds cling. Use a spoon to scrape the squash into a bowl, taking care to break up the squash into individual threads. To the squash, add the sautéed vegetables, sun-dried tomatoes, rosemary, and 3 tablespoons of the Parmesan cheese. Toss well to mix.

Place the mixture in a shallow baking dish; sprinkle with the remaining 2 tablespoons cheese. Bake until heated through, about 20 minutes. Serve hot.

Yield: 4 to 6 servings

Tomato-Stuffed Peppers

4 medium-size bell peppers
1 tablespoon olive oil
2 shallots, sliced
1 garlic clove, minced
1 cup chopped sun-dried tomatoes marinated in oil
1/2 cup cooked rice
1 tablespoon chopped fresh basil or chives or 1 teaspoon dried
1 teaspoon chopped fresh rosemary or 1/4 teaspoon dried
1/4 teaspoon salt
1/4 teaspoon pepper
2 tablespoons grated Parmesan cheese

Cut a thin slice from the tops of the stem ends of the peppers. Core and seed the peppers. Preheat the oven to 400°F.

Heat the oil in a small skillet. Add the shallots and garlic; sauté for 2 minutes. Add the tomatoes, rice, herbs, salt, and pepper; stir until well mixed. Spoon a portion of the mixture into each pepper shell. Place the peppers in a shallow baking pan. Sprinkle the cheese on top. Bake, uncovered, for 10 to 15 minutes. Serve at once.

Tomato-Stuffed Tomatoes. Substitute 6 medium-size ripe tomatoes for the peppers. Cut a thin slice from the tops of the tomatoes. Scoop out the pulp. Discard the seeds and chop the pulp. Proceed with the recipe above, adding the fresh tomato pulp with the dried tomatoes.

Yield: 4 servings

Roasted Potatoes

8 cups small new red potatoes (or quartered large red potatoes)
1/4 cup olive oil from marinated tomatoes
2 teaspoons garlic salt
1/2 cup chopped fresh basil or 2 teaspoons dried
1/2 cup diced sun-dried tomatoes marinated in oil

Boil the potatoes in water to cover until they are not quite done, about 4 minutes. Drain well and pat dry. Preheat the oven to 350°F.

On a baking sheet, sprinkle the potatoes with the oil. Roll the potatoes in the oil until well coated. Roast until brown and crisp, about 30 minutes.

Spoon the potatoes into a large serving bowl. Stir in the garlic salt, fresh basil, and tomatoes. Serve hot.

Yield: 6 servings

Specialty Cookbooks from The Crossing Press

Biscotti, Brownies, and Bars

By Terri Henry

This collection of easy-to-follow recipes presents more than 70 recipes for cookies baked in a pan.

$6.95 • Paper • ISBN 0-89594-901-6

Old World Breads

By Charel Scheele

In this authentic collection, the art of old world bread-making is available to everyone. Instructions are given to get brick oven results from an ordinary oven using a simple clay flower-pot saucer.

$6.95 • Paper • ISBN 0-89594-902-4

Pestos! Cooking with Herb Pastes

By Dorothy Rankin

"An inventive and tasteful collection—it makes the possibilities of herb pastes enticing."

—*Publishers Weekly*

$8.95 • Paper • ISBN 0-89594-180-5

Salad Dressings

By Teresa H. Burns

Making fresh salad dressings is easy! This handy little book is full of creative dressings that are fresh, healthy, and delicious. Teresa H. Burns has developed her recipes over years of entertaining.

$6.95 • Paper • ISBN 0-89594-895-8

Sauces for Pasta!

By K. Trabant with A. Chesman

"This little book has my favorite new and old sauces."

—Grace Kirschenbaum, *World of Cookbooks*

$8.95 • Paper • ISBN 0-89594-403-0

Salsas!

By Andrea Chesman

"Appeals to me because of the recipes' originality and far-ranging usefulness."

—Elliot Mackle, *Creative Loafing*

$8.95 • Paper • ISBN 0-89594-178-3